会呼吸的景观
BREATHING LANDSCAPE

水石设计 著

同济大学出版社

序言

李岚 / 水石设计创始合伙人

二十年前，我们几个初涉设计界的文艺青年，选择了景观中两个最传统、最经典的自然元素："水"和"石"，作为这家"景观环境设计公司"的名字，从此，踏上了求索和实践景观设计本质的漫漫不归路！

那个年代做景观设计的，"人法地，地法天，天法道，道法自然"——老子的这句话如同宝典最先根植于心！然而"自然"究竟是什么？貌似心中一直忽明忽暗；景观设计是否做到"虽为人作 宛自天开"就是最高境界？"人工"与"自然"的关系一直在纠结；大自然真的是自然的吗？……尽管一时并没有得到明确的答案，但是水石景观的小伙伴们从来就不会因此而停顿，一直坚持着边学习、边实践、边总结。

最近在看刘慈欣的《三体》，觉得非常有意思，常常会有一种脑洞大开的感觉！相比拯救地球、拯救全人类的大场景、大主题，一些细节反而更让人茅塞顿开。比如提到一个"地球演化数学模型"——它是用来模拟地球表面形态在过去和未来的演化。模拟的过程中，当把"生命"选项去掉，看看地球在没有生命的状态下演化到现在是什么样子，出人意料的是出现了一颗橙黄色的"干地球"，与其他的行星没有分别！所以书中说："地球产生了生命，生命也改变了地球，现在的地球环境，其实是两者互相作用的结果。……现在的地球，是生命为自己建的家园，与上帝没什么关系。"所以突然觉得，一切一切的根本就是"生命"！而人类作为生命的经典典范，和其他生命一起，塑造了我们的地球环境，形成了"自然"本身，可以说"人"即是"自然"之一。

也许，这可以帮助景观设计师跳出"人工与自然"之争，脱开那一种摇摆——是该去模法自然而隐在自然表象之下，还是该去彰显人类技术的强势，从而真正探索出景观设计乃至设计的本质。

"设计"——"设一个计谋",从字面上看就是非常具有人类特征的。有趣的是在《三体》中描写三体文明暂时对人类文明妥协时显示的一行字迹就是:"我们还是失败在计谋上。"所以说,"计谋"是人类文明宝贵的利器。

也许我们可以从如何让"设——计"这个人类行为"自然而然,成其所以然"来找寻设计的本质。无论塑造的景观最终呈现的是如何的性状面貌,设计的过程应该是回归融入自然的过程,尊重生命的过程:尊重她的使用者"人"——即我们经常强调的为人服务;尊重她本身——即赋予设计对象以生命感。

所以,非常喜欢这本书的名字:《会呼吸的景观》。景观会呼吸了,某种程度上就意味着她获得了生命。在这些大大小小的项目里,生命的张力呼之欲出:我们很容易地从长春水文化生态园改造再生后市民们朗朗的笑容里体会到这种生命的欢愉;也很明显地从红土遗址公园独特的肌理中感受到这种生命的演变……

获得了生命的景观设计正是我们水石人所孜孜以求的!

希望我们设计的景观能够自然而然,生生不息……

PREFACE

Lan LI / FOUNDING PARTNER OF SHUISHI

Twenty years ago, when we were still literary youths who first entered the design society, we chose and used two of the most traditional and classical natural elements in landscape, water and rock, as the name of this landscape and environment design company, and have ever since embarked on a long way to seek and practice the essence of landscape design.

At that time, landscape designers cherished in their heart Lao-tzu's teaching, "human learns from the earth, the earth learns from the heaven, the heaven learns from Tao, and Tao learns from the nature". However, what is exactly the "nature"? Sometimes, we seem to understand it, but sometimes not. Suppose we have achieved the effect that the "man-made appears just like the nature", is that the highest level of attainment in landscape design? The "man-made" and the "natural" have always been intertwined in relationship. Is the nature really natural? ……We did not have any definite answers at that time. However, we, people of W&R Landscape, never stopped our steps and persisted in learning our lessons in practice.

I'm now reading Liu Cixin's The *Three-Body* Problem and find it very interesting, even enlightening sometimes! Compared with the grand scenarios and great themes of saving the earth and mankind, some details are even more revealing. For example, the novel mentions a mathematical model for simulating the earth evolution, which simulates the evolution of the landform on the earth in the past and in the future. During the simulation, the option of "life" is turned off to see what the earth would be like today after all those years of evolution without life, and, unexpectedly, an orange-yellow "dry earth" appears, which is not quite different from any other planets! So the novel says: "the earth creates life, and the life changes the earth. The earth environment today is actually the result of the interaction between the two……The earth today is the home that life has built for itself and has nothing to do with god."

And I suddenly realize that "life" lies behind everything! Human being, as a classic example of life, has shaped our earth environment together with the other lives, and has formed the "nature" itself. We can say that "human" is part of the "nature".

Perhaps, this may help landscape designers forget the "man-made vs. the natural" debate and jump out of the swing between "shall we learn from the nature and hide under the natural surface or shall we exhibit the strength of human technology". Only there can we truly explore the nature of landscape design and even that of design.

Design literally means to plan something and make detailed drawings, a typical human activity. It is interesting to note that, in The *Three-Body* Problem, when the three-body civilization compromises temporarily to the human civilization, there displays a line saying: "we failed in planning after all". Therefore, "planning" is a precious weapon of human civilization.

Maybe we can find out the essence of design through letting this human behavior of planning "naturally create what is natural". No matter what the created landscape would be like finally in appearance, the process of design must be a process of returning to the nature and a process of respecting life: respecting "human", her user, because we often emphasize serving the people, and respecting herself, to give life to the object of design.

So, I like the title of this book very much: *Breathing Landscape*! When it breathes, a landscape, to some extent, has a life. In all these projects, large or small, life is right there: we can easily feel the pleasure of life in the bright smile of the local people after Changchun Culture of Water Ecology Park was renovated, and the evolution of life in the unique texture of the Red Earth Site Park……

That the landscape design comes to life is exactly what the W&R people are striving for!

I hope the landscape we design will be natural as the nature could always be……

人与物的情与境

张淞豪 / 合伙人 水石景观总经理

引

"水行石中，人穿洞底，巧逾生成，幻若鬼工，千溪万壑"，明代袁宏道于《袁中郎游记·园亭记略》有这样的描述，可见景观不仅是自然事物，更是一种自然的叙事，一种审美标准，是人在追求自然的极致状态。事实上，自然相对于人而存在，只要有人就没有纯粹的自然。人永远是在追求自然过程中消费自然。我们所倡导的自然，本质上是对自然的愧疚，是人在追求自然的过程——是人工自然。如同人工智能一样，它被后天重新定义，是一种相对概念。人工没有情感、缺少生命，但自然反之。所以我们必须将人工转化为对自然体悟的延续，增进身体与自然对话，促使人与自然共生。

景观是人与自然的交互媒介，通过景观的营造，试图对自然和再现作出思考，有关景观的发思，并不是发明设计交互程序或制定设计规则，也不是重建艺术形式和技术构造。而是把景观纳入一个精神所属的物理空间中，来触发人与自然的和谐关系。通过"会呼吸景观"的设计观念塑造场所，使人参与到自然中增进感情获得快感。换而言之，人们对于一处景致的沉浸实质上还是回到自身，是自身在体验生命活力的流淌，并由此获得精神上的愉悦。通过设计的空间让人、物、情、境高度关联。

人与境

从古至今人类对于回归自然的决心一直没有丢掉，在中国的文化里，景观并非花草树木、叠石流水这么简单。古代文人骚客一直有这样一种集会，借助自然天地作为宴集文会场所，幕天席地，各自找到一个自然物体让自己停下来，这个物体可以是石阶、树木亦或是林荫，配合自然物体以相应的身体姿态，停、倚、座、卧。在这种集会中，每个人都要找到自己的定位，依据地形配合姿势，依据地形决定交谈方式。当人的行为、精神与自然环境完全融合时，一副会呼吸的自然山林雅集图也就自然生成了。

上海徐汇区田林路街道改造项目，总长约 600 米，占地 1.1 公顷，周边社区基本是 20 世纪 50 年代左右建成，社区中缺少小孩子玩耍、交流；老人纳凉、晒太阳、跳广场舞的场所；街道店铺由五金日杂、水果店铺、奶茶餐饮、菜市场、学校邮局等构成。原有的街道没有让人与环境发生关联，让这么多活动需求参与其中，但是这些所有的日常活动都与街道密不可分。一个以日常社会交往为主的街道，是居民生活起居必备的场所，必须满足众多居民居住使用需求。

街道景观希望改造成满足各种人群活动串联，一个有人情味的场所。利用社区出入口绿地建立以兴趣点为功能分区的社交型口袋公园。使不同人群集合在一起攀谈、象棋、唱歌舞蹈；街道两侧安全护栏改造后倚靠着拎着菜的拄着拐杖的老人。街道端头的集中绿化带形成了一片共享的林下广场，在树林下有在交流的老人，有三五小聚的舞蹈爱好者，小朋友滑着滑板车穿梭于其中。改造后的街道与周边社区人群紧密关联，充分利用现有空间（街角、绿地、街道两侧）为日常行为的发生提供场景，通过街道将各种人群和活动场地融合在一起，重新描绘了一幅活色生香的梧桐树下社区街道雅集图。

物与情

在追求自然方面造诣颇丰的日本也不例外，它们的"诧寂"美学，强调自然朴素空寂自然之美，这种美针对人、自然、物乃至社会世相，感物生情，物心合一。在日本文化的视野里，它崇尚自然，取材自然，甚至象征自然，它没有太多人工雕琢的痕迹，而是创造出一种返璞归真、清宁极简的禅意境界，强调超脱自然。

长春水文化生态园项目原址是一座建于 1932 年的水厂，改造设计中最大化保留了原始自然生态环境，最大限度保留场地记忆；利用原有道路、场地植入功能空间，将功能与自然环境融合，材料选择以就地取材为主，场地中的枯树被机器打碎铺设在森林中，形成天然松软透水路面。人行走在其中，如同踏入原始森林，废弃的枕木改造利用为地面铺装、树穴、座椅，与乔木形成舒适的具有时代记忆感的林下休憩空间；场地中原始的石料被重新铺装到广场上，与时间一起被无数市民来回打磨，迎着落日余晖的掩映，仿佛能感受到 80 多年前的场景及厚重的场地记忆。

营造景观并不是为了模仿自然，而是为了赋予自然与情感及生命。日本庭院"作庭者"会把自己置身于不同空间，用最质朴的材料，甚至会花上几个月精心打磨每一粒石子。通过在长时间的打磨与付出中感受自然、感受朴素简练、超然自在艺术境界。

水厂项目改造中功能场景的定位如同"作庭者"打磨沙子一样，设计师对场地充分观察与认知，对四季季相、昼夜变化、日月情境进行调研，感受时间轨迹和自然的变化。除此之外对每一个可改造区域进行观察，分析光照、现场自然条件、景观资源等，多维度比对后确定功能定位、判断位置朝向及甄别选址。设计中尽量减少土地开发对环境的影响，采用低干预的设计手段，实现自然生态环境的可持续性。建成后的水文化生态园，走在空中栈桥上，阳光从树梢斜射下来，透过树枝，光影的变化产生丰富的效果。不同角度的光晕，仿佛时间在飞逝，四季在森林中流动。顺着螺旋楼梯而上，直达树屋顶部赤足盘坐，眼前是郁郁葱葱的森林、树下是一片清澈的水池还有两只鸳鸯在戏水。闭目冥思，微风拂过，人与自然充分融合，呼吸共生。

"会呼吸景观"是一种人与自然交互的精神化物态空间营造，由外到内营造出一个交互体验性场所，它所带给人的并不是尊贵和仪式感，而是有深远延伸的空间感。是强调人、物、情、境在物态空间中的高度关联。会呼吸景观对"外"是自然交融，对"内"是与人交互，是人身体、精神的内外呼吸。会呼吸的景观是一个不可度量的自然属性，它能引人入胜，引导人进入凝神思考与宁静的自我存在。更重要的是，它不止关注人与自然，而是将人、自然、社会世相的叠合，是人的生命流淌过程中在空间环境中的体验。

其实世间本无景观，因人而有景观，无论你身居何地，景观都与你息息相关。会呼吸景观希望给人营造出舒适的、幸福的、会呼吸的自然情感交互状态，这种情感化的景观营造方式，融合在本书的角角落落，让你在不同的角落里自然呼吸。会呼吸景观是近年来水石景观设计普适性的价值观、设计观、哲学观。这套价值体系的诠释需长时间项目实践与佐证，同时也需要市场行业来审视。水石设计也希望假以时日，集腋成裘，为景观行业的价值体系贡献绵薄之力。

FEELINGS OF PEOPLE AND THINGS IN THE ENVIRONMENT

Songhao ZHANG / GENERAL MANAGER OF SHUISHI LANDSCAPE

Prologue:
In the Ming Dynasty, Yuan Hongdao described in the "Travel notes of Yuan Zhonglang on gardens and pavilions": "Water flows between rocks and people pass under caves. The man-made landscape looks natural, as if created by supernatural beings. There are so many creeks and gullies that ......" Therefore, landscapes are more than just natural things. They are a narration of the nature, a criterion of aesthetics and an ultimate state of human pursuit of being natural. In fact, the nature is there alongside human beings. Where there are human beings, there is no pure nature. Man consumes the nature in his pursuit of being natural. Our advocation of being natural is essentially a feeling of guilt about the nature as well as a process in which human being pursues the nature - an artificial nature. It is, like the artificial intelligence, redefined after its creation and is a relative concept. Artificial creations have no emotions or life, but the nature has. Therefore, we must extend artificial creations as per our understandings of the nature, enhance the dialogue between human body and the nature, and promote the symbiosis of man and nature.

Landscapes are an interactive medium between man and nature. We try to think about the nature and the reproduction of nature through the construction of landscapes, and our thinking about landscapes is not for the invention and design of interactive programs, formulation of design rules or reconstruction of art forms and technical structures. It is, on the contrary, bringing the landscape into a physical space where the spirit belongs and establishing a harmonious relationship between man and nature. In a place constructed with the design concept of " Breathing Landscape", people, being part of the nature, improve their feelings with pleasure. In other words, people's immersion in a scene is actually a return to themselves, their experience of the flow of life, and the enjoyment of spiritual pleasure. The designed space has people, things, feelings and environment correlated.

People and environment:
Since ancient times, human beings have never lost their determination to return to the nature. In the Chinese culture, the landscape is more than just flowers, trees, rocks and flowing water. Ancient literary people always had gatherings at wild places as the venues for feasts and literary meetings, where participants found natural objects between heaven and earth to rest their bodies, which could be stone steps, trees or shades under trees, where they

comfortably toke such poses such standing, leaning, being seated or lying. In such gatherings, each person had to find his own position and adjust his posture and mode of conversation according to the terrain. When people's behaviors, spirit and the natural environment were fully integrated, a natural picture of the mountain and the forest gathering that could breathe came naturally into being.

The street reconstruction of Tianlin Road in Xuhui District, Shanghai, is about 600m long and covers an area of 11,000m2. The communities there were basically built in the 1950s, not much space for children to play or for old people to enjoy the sunlight or dances. Street shops were mostly hardware shops, fruit shops, milk tea bars and restaurants, grocery market, schools and post office, etc. The streets did not allow much connection between people and environment or had so many activities integrated in them. However, all the daily activities could not be separated from the streets. A street for daily social intercourse is a must-have place for the residents, and should meet the residential and other needs of the local people. Through streetscape renovation, the street is expected to meet the needs to interconnect various mass activities and become a place where people feel comfortable in it. The green space at the entrance to the community is used for a social pocket park with different functional areas of interest. People may gather in their areas of interest to chat, play chess, sing and dance; the guardrails along the street are transformed so that old people carrying canes or vegetables may rest against. The green belts at the ends of the street are transformed into public squares, where old people communicate under trees, a couple of dance enthusiasts exchange with each other, and children skate through the square. After reconstruction, the streets are closely connected with the people from the neighboring communities, and street corners, green spaces and pavements of the streets provide spaces for daily activities. The street combines various groups of people and places of activities, painting a quiet and graceful picture of vivid community life under phoenix trees.

Things and feelings:
Japan, a country highly accomplished in the pursuit of nature, is no exception. Their "Wabi-sabi" aesthetics emphasizes the beauty of natural simplicity and quietness. This beauty is found in man, nature, things and even the world, where feelings are generated in things and things

unite mentality. The Japanese culture advocates, learns and even symbolizes the nature, and does not have much human contrivance. Instead, it creates a Zen realm, returning to the nature, the quiet and the simple, and emphasizing detachment and being natural.

The original site of Changchun Culture of Water Ecology Park was a water plant built in 1932. In the reconstruction design, the primitive natural ecological environment and the memory of the site are retained to the maximum; on the basis of the original roads and sites, functional spaces are designed to integrate functions with the natural environment; materials are taken from the site, and dry trunks are crushed and laid in the forest to serve as a naturally soft and permeable pavement. Walking in the park, you feel like stepping into a primitive forest. Used crossties are transformed to pave the ground, build treehouses and make chairs, creating a comfortable space for rest under trees with a memory of the past times. Old stones at the site are reused on the square to polish with time by countless citizens walking on them, reflecting the setting sun that disappears every now and then, and you can feel the scenery and deep memory of the site over 80 years ago.

Landscapes are not created to imitate the nature, but to endow the nature with feelings and life. A Japanese courtyard builder usually places himself in different spaces, uses the most primitive materials, and even takes months to grind each pebble carefully. In the long time of grinding and giving, he feels the nature, perceives the simple and concise, and stays detached from the world and emerged in the artistic realm.

In the water plant reconstruction project, the designers design the functional scenery as the Japanese courtyard builder grinds pebbles. They observe and understand the site, study the changes of seasons, days and nights, the sun and the moon, and feel the evolution of time and nature. Besides, they observe every area to reconstruct, analyze the lighting, natural conditions and landscape resources, and determine the functions, locations, directions, and the selection of sites after comparison from various perspectives. In the design process, every effort is made to minimize the impact of land development on environment, and low intervention design method is adopted to realize the sustainability of the natural ecological environment. On the elevated footpath in the completed water culture ecology park, the slanting sunlight shoots down from treetops, and the changing light and shadow in the branches produce a wonderful effect. Halos in different angles give people the impression that time fleets, and four seasons flow in the forest...Climbing up the spiral staircase to the top of the treehouse, taking off your shoes and hunkering down, you will find nothing but luxuriant forest, crystal-clear water under the tree and two mandarin ducks playing in the pond. Close your eyes and meditate, and let the breeze embrace your face, experiencing a harmonious coexistence and a full integration of man and nature.

" Breathing Landscape" is a spiritual creation of physical space for man and nature interaction, a place of interactive experience from the external to the internal. It brings about not dignity or ceremony, but a sense of space that has profound extension. The emphasis is on the correlation of people, things, feelings and environment in the physical space. For Breathing Landscape, the "external" is the natural integration and the "internal" is the interaction with people, and is the internal and external breathing of human body and spirit. Breathing Landscape is an immeasurable property of the nature, which is fascinating and lead to meditative and peaceful self-being. More importantly, it focuses not only on man and nature, but also on the superimposition of man, nature and the world, the experience of flowing human life in the spatial environment.

In fact, there will be no landscape in the world unless there is human being. No matter where you live, you are closely related to landscape. In "Breathing Landscape ", we hope to create a comfortable, pleasing and breathable interaction of the nature and feelings. This emotional landscape creation method is reflected throughout the whole book " Breathing Landscape" is a collection of the universal values, design concepts and philosophies that SHUISHI has employed in recent years. The interpretation of these values needs long-term project practice and supports, as well as the examination of the market and the industry. It is expected that SHUISHI will gather more experience over time and contribute to the values of the landscape industry.

景观三重门

石力 / 合伙人 水石景观设计总监

景观是什么？

从学习景观到从事职业的景观设计已经近20年了，却觉得自己还是一个"景观的小学生"，以至于到现在我还不能描述出景观的准确定义，景观是什么？景观做什么？为什么做景观？接触景观越久，越觉得它包含的内涵越丰富，它的边界越混沌，中国景观发展这十年在数量上有巨大的发展和积累，在质量上又有多少遗憾和思考，促使我们内心对景观的思辨一直不停，有时甚至会清楚地意识到景观不该是什么，景观设计应该怎样，景观要是能这样就好了……凡此种种，很多思绪挥之不去，记录下来是对自己头脑的梳理，也是对自我的批评、提醒和鼓励，如果能和广大景观人一起共享交流该是一件值得高兴的事。

伪美学的景观

景观不只眼前的一亮！
近几年的设计工作中接收到最多的设计要求是要"眼前一亮"，特别是住宅设计领域大量的设计资源投入在营销展示区这一销售道具的设计上，钟情于"眼前一亮"的表面视觉化设计观，往往只在设计一种样式、风格、情境，景观场地如展示道具般在销售活动结束后被拆除荡然无存，这几年间已经把教科书上的各种景观风格做完，从新古典、新中式到现代简约去风格化，近期更是"景观创新"迭出，各种新颖样式层出不穷，我们也不乏融入这种洪流并贡献自己的力量。但是在这样的景观中能够体会的东西却非常有限，感觉景观设计离我们生活中的真实感受越来越远，我们堕入到一种唯表面化的设计观中，颜值即正义？在这里必须强调设计的形式美是非常之重要的，但是只是片面追求外在形式就有失景观对真美的承载。
景观设计应该追求什么？什么是好的景观设计？不禁扪心自问。当下的社会是一个物化的社会，似乎所有的东西都可以变成商品购买，食物、衣服、汽车、建筑都可买，不需要的东西可以创造需求购买，买买买。景观可以被购买吗？当然也是可以的，景观的场地可以被购买，但是景观却能创造不能被购买的东西，空气、阳光、水的体验，人与环境的关系，人与人的关系，人与自己的关系，人与其他物种的关系，这些或许是我们应当关注在表面形式之后真正的美，也是作为景观设计师的幸福和责任所在。

景观的目标是去伪存真，大胆畅想景观设计的三重门，也是三重境界：一、自然意想；二、会呼吸的关系；三、从共享到共生。

第一重门：自然意想

景观中的自然是什么？自然不是种树！
常收到这样的设计要求"这个景观不够自然，要多种树"，"绿植再多点就自然了"这个理论一直以来是笼罩在景观设计师心头的一片阴影，我们曾经历过把一处场地种满绿植也没有达到自然的感觉，景观中的自然到底是什么？可能真正的自然在人类社会一产生就消失了，自然是一种遥远的存在，够不着，摸不到，但时刻存在，在意念中时时勾着你、诱惑你、威慑你，自然在彼岸，得不到它，真实的自然是有它可怕的一面的，在人类社会发展之前的状态，有野兽、天会黑、发洪水，我们说的自然肯定不是这一面，显然是大自然对人类友善的一面，是自然中的真和善。

那我们说的设计中的自然是什么？最浅显的，是自然感，自然而然，自然美好的样子，松弛的、真诚的、博爱的感觉。在设计里是让人想到一种山、河、草木的感觉就是这样，这种感觉是可以放大的，一个盆景就能达到这个效果，一院枯山水甚至不种一棵树也能放大"真"的感觉，即使枯山水是"假的"，设计就是要放大这种感觉，手法不一定是"真的"但情感是真的。我们在很多优秀的建筑中也见识过这种空无一树却无限自然的体验。在"昆明樾府"的设计中用就是自然意想的手法，把云南的田、人、屋、山和谐相融的状态抽象浓缩在一方现代庭院之中，用大地艺术的手法让地景和建筑连为一体，地面景观用民居中屋瓦的元素层层叠叠而出，当人被置身于一片屋瓦的海洋中，看到大地和建筑连为一体时似乎感受到一种似曾相识的自然意向。三亚万豪酒店景观在临海的镜面水上创造了一处艺术构架，构架的设计来源于海风的形态，仿佛凝固了的海风，本来静止的构架也具有了动态的感觉，在这里抽象的海被转化成变幻的光、影、风、形，在人的坐、卧、行间仿佛重新看到了大海的样子。

景观在这个层面是希望创造一处场所和具体的物件，通过它产生对"遥远自然"的感情和意想。这个是景观破解自然的一种思想和手段，也是景观设计的第一重境界。

第二重门：会呼吸的关系

景观不只盆景和枯山水！
虽然景观要通过对自然的意想去创造和抽象，但景观却仍然不是对自然形式的简单拷贝、模拟，其实也拷贝不了。
景观更重要的是建立、酝酿、培育一种关系。

会呼吸是一种关系，也是一种态度。没有呼吸，代表死亡，没有呼吸的景观就是死去的景观，呼吸是自然的普遍状态，呼和吸是相反的动作，却是对立统一的。

会呼吸是建立一种空间关系，空间是有阴阳的，阴阳相生，呼吸是打通、建立、创造关系和联系，良好的联系。福州云影子会所的设计是在尝试通过创造一个"峡谷"空间楔入地下，将光线、空气引入建筑的地下部分，同时"峡谷"长长的空间的空气对流形成了风，峡谷的高差结合水流创造了瀑布，建筑的正形态和峡谷的负形态形成一种互为因果的空间关系。

景观中的自然是符合自然规律、顺应自然规律，自然就是呼吸。景观最重要的是建立和阳光、空气、风……的关系，自然和人工是对立面，就像人们往往描述"这个设计太人工了"把自然和人工放在了对立面，而又不是对立面，人工可以创造自然的感觉，和自然环境建立良好的联系，创造很好的接受阳光、空气、风的关系，"人的工"本质也在自然的概念里。三亚万豪酒店景观设计中，我们设想如果身在海边，不能感受到海风的吹拂，如果视线不能延续直到海面，这样的场所就不具有临海的特质，我们就需要重新找到和大海对话的方式。这是一个酒店因品牌升级而改造外环境的景观设计，原场地虽临海但现状空间形态没有形成和海景的联系，植物甚至阻隔了看海的视线，为了让人和海形成互动，设计以分析海边的风、光环境为基础，创造了顺应海边自然要素的场所形态和要素。

在景观中建立和自然气候因素的联系是一种会呼吸的景观方式，利用土地、风、光、空气、水、这些气候因素的自然规律创造会呼吸的景观，能让我们回归自然的感知，人们更愿意更长时间在室外景观中停留感受是我们的目的，微气候设计是顺应气候而非反抗气候的态度，微气候不光是满足气候条件，而是追求气候和美学的完美结合。

会呼吸是交流，建立人和人的联系，孕育一种存在的状态。在临沂星河城生命之树项目中，引入"大树"的概念，因为"大树"除了树本身也创造了人与人的关系场所，回忆小时候庭院的大树下边，总是吊着一个摇荡不停的秋千，那一根细细的绳子给了我们无尽的欢乐，坐在秋千上看大树结出果实，看树叶随风飘落，夏天我们沐浴在大树的光影之下，晚上看着天上的星星听奶奶讲故事，一棵树就是一个家的空间，围绕大树建立了人的关系。如果，我们创造一种空间，让它像大树一样存在。当构架能够如树冠一样遮阴和透光，等大树结出果实孩子可以在上边玩耍，生活就可以在大树下生长。因为如"大树"一般的构架我们能和家人欢聚，在这能结交更多的朋友，那么没有真树也有自然。

在景观中，每种关系都不会重复，每个设计都有独特的条件，所以每个设计都应挖掘和建立各不相同的关系，每个设计都是独特的、唯一的。
会呼吸是景观的活化，是景观设计的第二重境界。

第三重门：从共享到共生

景观不是小桥和流水！
景观经常被看作是模拟一种情景，曾经的"风景园林""环境艺术"都已不能涵盖景观的内涵，这些称谓显得非常局限。就像前面提到景观能创造不能被商品化的东西，人与环境的关系，人与人的关系，人与自己的关系，人与其他物种的关系。

历史上的欧洲古典园林、中国苏州园林固然是包含了人类文化艺术的精髓，但很大程度上是皇家贵族和高级官宦文人的独享空间，随着城市发展带来思辨，在城市中私享的价值受到挑战，我们不断反思在城市中生活和存在的状态，什么样的关系会带给城市更多的活力，开放共享开始变成人们优秀的理念和未来，景观更大的胸怀和前景或许是共享。共享是一种关系，景观通过物质空间的创造，能否酝酿和培育共享的、人与人相处的状态和自身存在的状态？田林路改造项目中，通过观察分析街上人群活动，带来对街道和城市空间的重新思考，创造共享街道空间进而孕育出丰富的人与人的关系，重新赋予一条老旧的街道以活力。

人和土地是怎样的关系，景观连接着土地的过去、现在和未来，一处原生态的环境当景观介入时，是强力的改造它还是慢慢地融入，是否无为的设计胜过有为。

在安南小镇红土遗址公园，我们初次到达场地，看到一片如波浪起伏的自然红土地貌，地质活动造就奇幻的自然奇观，我们的第一直觉是这里不需要设计，走在碎石小路上，穿行在茅草丛间已经非常惬意，这里不设计就已经很有体验，感觉可以"无为"，但是城市开发还是势必要把这里变成一座公园，景观设计就以最低的介入为原则，尽可能保护原生的形态并恢复其更生态的状态。一切景观构筑都用最轻的介入，仿佛在原生的地貌上浮动飘起，我们希望当未来公园开放游人感受到的是和我们初次看到的一样的红土地、小径和茅草。这是当"不得不为"时是否可以"有所为有所不为"。

一直以来设计强调"人性化",以人的利益为核心,我们有时会认识到世界是一个循环,人是一个复杂链条中的一环,人的状态会受到身边其他人、环境、植物、动物的影响。我们介入一块土地一定会干扰到这里现存的动物植物,人不是孤立的存在,当景观能成全其他物种的生存时,在这个大循环中最终受益的还是人,通过景观人的生活是可以和各种关系取得平衡的,这个共生的状态让景观设计就进入了一个全新的层次。长春水文化生态园原是城市中一座废弃的水厂,人类生产活动的退出让植物和小动物找到一处城市喧嚣外的栖息地,甚至在这里茁壮生长了,初到场地时被城市中还能有这么一片参天大树震撼,密林的状态让人几乎不能穿行其中,我们决心在人的城市公共活动需求和已有的植物繁盛、动物栖息间找一个平衡,林间的道路全部架空穿行,栈道基础设计成最精简的构造,小路全部采用不要基础的形式:枕木、枯树皮、砾石达到最小干扰的效果。在公园中任其保留大量的树林不去打扰,在对原有建筑的改造中也着重保护长年攀缘在建筑外的爬藤植物,在这里物质空间已经和动植物连成了一个整体的存在。在共生中我们看到了一种美,和发自内心的平和和幸福。从共享到共生是景观设计的第三层境界。

走出三重门

"三重门"原指中国古代衙门的三重门坎,隐喻古代衙门官府登记森严的层层围合,景观的内涵从风景、园林、环境艺术,到"景观都市主义",概念本身就像一个个门坎划定界限的同时也限制了思维,我们思考景观的三重门坎,"自然意想""呼吸关系""从共享到共生",虽然三种境界逐级提升,跨过一重门有时又进入另一重门,到最后是希望能走出这三重门让景观设计还复到一个没有束缚的自由世界,这是对自己的思考、批评和勉励。

THE TRIPLE GATES TO LANDSCAPE

Li SHI / DESIGN DIRECTOR OF SHUISHI LANDSCAPE

What is landscape?
About 20 years ago, I began to study landscape and then became a professional landscape designer. However, I still feel like a "pupil" today and cannot define exactly what landscape is. What does landscape do? Why do we do landscape? The longer you are immersed in the landscape business, the richer her connotation and the more ambiguous her border you will find. The Chinese landscape business over the past ten years has achieved enormous development in quantities. However, as enormous are the regrets and meditations on quality, we're urged to think further about landscape. Sometimes we realize clearly what does not belong to landscape, what should landscape design be like, and it would be wonderful if we could do landscape in this way...All these thoughts have lingered in my mind so long that I would like take them down to sort out my thinking as well as to criticize, remind and encourage myself. It would be wonderful if I could share them with my fellow landscape professionals.

Landscapes of pseudo aesthetics:
Landscape is more than eye-catching!
In recent years, the most common design requirements that we have received in our work are "eye-catching", especially in the field of residence designs, where large amounts of design resources are invested in the design of display areas for marketing use. The clients are fond of superficial visual designs that "catch the eye", and are addicted to a specific pattern, style and situation, and the site of the landscape, together with the display props, are demolished after the marketing activities are concluded. In the past few years, all the styles of landscapes in the textbook, from new classics, new Chinese styles through modern style-free simplism, are used up. And in the recent "landscape innovation", numerous novel patterns are produced, and in this torrential flood we are also involved, and have made our contribution. In such landscapes, however, we can feel little. The landscape design increasingly alienates us from real feelings in life. We fall into a kind of superficial design concept, where beautiful appearance is everything. I must admit that the beauty of form is very important. However, when the external form becomes the only pursuit in landscape design, the landscape will be unable to carry true beauty.
What should we pursue in landscape design? What is a good landscape design? I can't help asking myself. The society today is a material one. It seems that everything can be

purchased, food, clothes, cars, buildings, you name it. When something is not needed, the demand can be created. So, just buy it. Can landscapes be purchased ? Of course. The site of the landscape can be purchased. However, the landscape can create things that can not be purchased, e.g., the experience of air, sunshine and water, the relationship between man and environment, the relationship between people, between people and themselves, between people and other species of life. These may be the true beauty under the superficial form to which we should pay more attention, and there lies the happiness and responsibility of landscape designer.

The goal of landscape is to remove the false and preserve the true. There are three gates or realms to landscape design: 1. imagination of the nature; 2. breathing relationship; 3. from sharing to symbiosis.

First gate: imagination of the nature

What is the nature in landscape ? The nature is more than planting trees!

We often receive such design instructions as "this landscape is not natural enough: We need more trees", and "it will be natural when you place more green plants there". This has always been a theory that frustrates landscape designers. We have had the experience that the whole site is planted with green plants without achieving the feeling of being natural. Perhaps the real nature has disappeared the moment the human society comes into being. The nature is a distant existence that cannot be touched. However, it does exist in your mind, hooking you, tempting you and threatening you all the time. The nature is on the other side and you cannot get it. The true nature has its terrible side. Before the human society began to develop, there were wild animals, dark sky and floods. This is definitely not the side of the nature that we are talking about. Obviously, we are talking about the nature that is friendly to human being, the genuine and kind side of the nature.

So, what do we mean by the nature in design ? First, the concrete nature, including mountains, rivers, trees and our indispensable sunshine, air, water, so on and so forth. The second is the abstract nature.

an artistic structure is created on the mirror water by the sea. The design of the structure originates in sea breeze, the form of which gives the still structure a dynamic feeling. The abstract sea here is transformed into changing lights, shadows, wind and shapes, and guests sit, lie and walk about as if in front of the sea.

At this level, the landscape is to create a place and a concrete object, through which feelings and imaginations of "a distant nature" are generated. This is a thinking and abstracting method to decode the nature by landscape, the first realm of landscape design.

Second gate: breathing relationship

Landscape is more than bonsai and dry landscapes!

Landscape needs the imagination of the nature for creation and abstraction. However, it's not a simple copy or simulation of natural forms. In fact, they cannot be copied.

Landscape is, more importantly, to establish, conceive and cultivate a relationship.

Breathing is a relationship as well as an attitude. The lack of breathing means death, so a landscape that cannot breathe is dead. Breathing is the common state of the nature. Inhaling and exhaling are opposite but unified in actions.

Breathing establishes a spatial relationship, and in the space, yin and yang are born together and coexist. By breathing, a relation, a connection, a nice connection, is opened up, established, created. In Fuzhou Peninsula Cloud, the designer attempts to create a "valley" to wedge underground to introduce light and air into the basement, and, at the same time, the air in the long valley circulates to create wind, and water flows in cascade in the valley to establish a causal relationship between the positive building and the negative valley.

The nature in landscape means following and conforming to the law of nature, and the nature breathes. The most important thing about landscape is to establish a relation with the sun, the air, the wind...The natural and the artificial are opposite to each other. When people say "the design is too artificial", they put the natural and the artificial on opposite sides. However, they are not so much opposite. The artificial can create a feeling of nature, establish with the natural environment a nice connection and build good relationships to receive sunlight, the air and wind. The essence of the "man-made" is also in the concept of nature. In the landscape design of Sanya Marriott Hotel, we imagine ourselves staying in a hotel by the sea. If the ocean cannot be seen and the breeze from the sea cannot be felt, such a place does not feel coastal, and we need to find a way to talk to the sea. This hotel needs to upgrade its brand and renovate the external environment, thus the landscape design. The site is close to the sea but the spatial formation offers no connection with the sea view, which is cut off by plants. In order to establish an interaction between man and the sea, the designer analyzes the sea breeze and lighting environment and creates space forms and elements in conformity with the seaside natural elements.

Establishing a connection with natural and climatic elements in landscape is a breathing way of landscaping. By making use of the natural laws of land, wind, light, air, water and other climatic elements to create a landscape that breathes, we can feel we are returning to the nature, and people are willing to stay outdoor in the landscape for a longer time. The purpose of microclimatic design is to suit, but fight against the climate. Microclimate will not only satisfy the climatic conditions, but also pursue a perfect combination of climate and aesthetics.

Breathing is communication, establishing relationships between people, and cultivating a state of being. In the Tree of Life project of Linyi Galaxy City, the concept of "big tree" is introduced. The big tree, apart from the tree itself, also creates a place for relationship between people. In our childhood memory, there is always a swing under the big tree in the courtyard, and the thin ropes are the source of our happiness. Sitting on the swing bench, we admire fruits riping in the tree and leaves falling in the wind. In summer, we bathe in the light and shadow of the big tree, and at night, we watch the stars in the sky and listen to Grandma's stories. The tree is the space for a family, and people's relationships are established around the big tree. We create a space, and let it exist like a tree. When a structure can create shades and let down some light like the crown of a tree, we can imagine fruits grow in big trees, kids play on the top, and life grows under the tree. If, because of the "big-tree" structure, we can unite with our family and make more friends here, we are already in possession of the nature even if the tree is not real.

In landscape, no relationship is repeated, and each design is unique in conditions. Every design explores and establishes a different relationship, and every design is unique.

Breathing embodies the life of landscape, and is the second realm of landscape design.

Third gate: from sharing to symbiosis

Landscape is more than small bridges and flowing water!

Landscape is often seen as a simulation of the scenery. The past concepts of "gardening" and "environmental art" are very limited and could no longer cover the definition of landscape. As mentioned earlier, landscapes can create things that cannot be sold, e.g., the relationship between man and environment, the relationship between people, between people and themselves, between people and other species of life.

Classical European gardens and Chinese Suzhou gardens are no doubt the essence of human culture and art, but they are largely private spaces for royal families, high-ranking officials and scholar. The development of cities challenges the values of private enjoyment, and we constantly reflect on the state of life and existence in the city: what kind of relationship will bring more vitality to the city？ Being open and sharing becomes an excellent concept and future, and a broader mind and prospect for landscape might be sharing. Sharing is a relationship. Can a landscape, by the creation of physical space, conceive and cultivate a sharing state among people and its own state of existence？ In the Tianlin Road renovation project, the designer observes and analyzes people's activities in the street, rethinks about the street and city spaces, and creates a shared street space to breed more people-to-people relationships, revitalizing an old street.

What is the relationship between man and soil？ The landscape connects the past, the present and the future of the land. When a landscape lands in a primitive ecological environment, should it be a dramatic transformation, or an imperceptible integration？ Isn't an inaction in design better than intervention？

When we first arrived at the site of Nanchang Red Earth Site Park, we saw a rolling red earth topography, a fantastic natural wonder of geological activities. In our intuition, there was no need for design. A walk on gravel paths through wild grass was very comfortable. Without any design, the experience was already excellent and all we needed was "inaction". Urban development, however, will turn this place into a park, and the landscape design is, on the principle of minimum intervention, to protect the primitive form and restore its ecological state as far as we can. With the slightest intervention in landscape construction, as if floating on the original topography, we hope the first visitors to the future park will feel as we did to the same red earth, narrow paths and wild grass. When we "have to", can't we just "do something and then refrain from doing more"？

"Humanism", to be human-centric, has always been the emphasis of design. Sometimes we may realize the world is a cycle and human beings are a link in this complex chain. Our state is subject to the impact of other people, the environment, plants and animals around us. When we come to a place, we are sure to disrupt the animals and plants living there. Human is not an isolated being. When a landscape helps the survival of other species of life, the ultimate beneficiary in this big cycle is human being. Through landscape, people's life can achieve a balance in various relationships. This symbiotic state pushes landscape design up to a new level. Changchun Culture of Water Ecology Park is on the site of an abandoned water plant of the city. When the production activities ceased, plants and small animals found here a habitat and thrived outside the hustle and bustle of the city life. When we first arrived at the site, we were shocked to find such huge trees in the city, and the dense forest made it impossible to walk through it. We were determined to find a balance between the public activities of the city dwellers and the habitat of flourishing plants and animals. The roads in the forest are all elevated, footpaths adopt the simplest structure in foundation, and all trails are designed without foundation: crossties, dry barks and gravels The stones and timbers are piled up with gaps for small animals to live in. Large plots of woods are retained and left undisturbed in the park. In the reconstruction of the original buildings, the climbing plants that have been growing for decades on outside walls of the buildings are also protected. The physical space here is integrated with wild life. We find beauty in symbiosis, and peace and happiness from the depth of our heart. From sharing to symbiosis, this is the third realm of landscape design.

Walking out of the three gates
The "triple gates" originally refers to the three layers of gates to the ancient Chinese government building, and stands for the strict registration formalities that the public have to undergo to approach the government. The connotation of landscape from scenery, gardening, environmental art to"landscape urbanism", so far as the concept is concerned, seems marked out by gates. So it restricts the thinking by such demarcations. When we think about the triple gates of landscape, "imagination of the nature", "breathing relationship" and "from sharing to symbiosis", our realms rise gradually as we progress from one gate to another. However, we hope, in the end, to forget about the triple gates and conduct landscape design in a free state. This state of freedom ultimately points to how to deal with the relationship between man and nature. The understanding of nature changes constantly with the evolution of social values.

Thoughtful architects have proposed the concepts of "defeated architecture" and "modest architecture", and begun to think about free and more harmonious states of architecture, e.g., dissipation, weakening, transparency, sharing, so on and so forth. All these implicitly point to the connotation of landscape. Can we say that the future of design is also the future of landscape？ Perhaps the era of landscape has already come……This is self-examination, criticism and encouragement.

目录
CONTENTS

序   2
Preface

人与物的情与境   6
Feelings of people and things in the environment

景观三重门   14
The triple gates to landscape

01   百年之技，千年之森   / 长春水文化生态园   28
Technology Of Several Hundred Years And Forest Of Several Thousand Years/Culture Of Water Ecology Park ChangChun

02   树梢上的漫步道   / 西安白鹿原西坡树梢漫步栈道   142
Rambling Road At The Treetop/Rambling Plank Road At The Treetop Of Western Slopes Of Bailuyuan XiAn

03   城市的微更新   / 上海田林路改造   170
Small Update Of The City/ ShangHaiTianlin Road Renovation

04   公园里的游憩时光   / 西安理想城   192
Tour Time in The Park/ Ideal City XiAn

05   城市里的公园盒子   / 南昌博览城绿带公园   212
Park Box In The City/Guobo Green Park In NanChang

06   上升的螺旋引力   / 南昌博览城庆典公园   226
Rising Spiral Gravitation/Celebration Park In NanChang

07   浮动的森林   / 南昌安南小镇玻璃花房   242
Floating Forest/ Annan Town Glass Greenhouse Nanchang

08	大地美术馆　　/ 南昌红土遗址公园	254
	Earth Art Museum/ Red Earth Heritage Park Nanchang	
09	寻找和大海对话的空间语境　　/ 三亚 JW 万豪酒店	274
	Find The Spatial Context Of Dialogue With The Sea/JW Marriott Hotel Sanya	
10	前卫与本真的对话　　/ 郑州长安古寨	288
	Avant-Garde And Genuine Dialogue/Zhengzhou Chang'an Ancient Village	
11	如大树一样存在的景观　　/ 临沂生命织树	302
	Existing Landscape Like The Tree/ Tree of Life In Linyi	
12	地域文化的现代回归　　/ 昆明樾府	316
	Modern Return Of Regional Culture/ Kunming Oriental Mansion	
13	景观的微气候综合体　　/ 福州云影半岛	336
	The Microclimate Complex Of The Landscape/Fuzhou Peninsula Cloud Club	
14	微气候的复合设计　　/ 福州溪溪里	356
	The Composite Design Of Microclimate/FuZhou Xixili	
后记	让景观更好地为人服务	368
	Postscript Let the landscape design serve people better	

百年之技，千年之森 | 长春水文化生态园
Technology Of Several Hundred Years And Forest Of Several Thousand Years
Culture Of Water Ecology Park ChangChun

地点｜吉林省长春市
规模｜34万平方米
业主｜长春城投建设投资（集团）有限公司

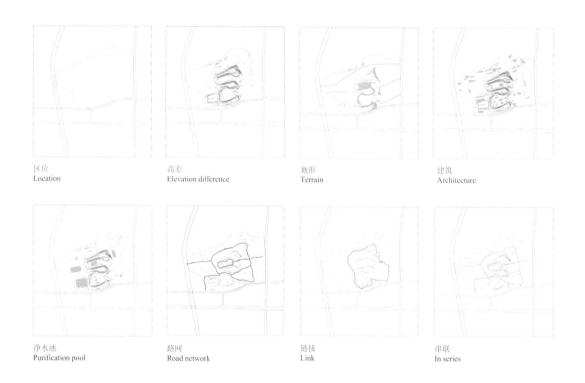

区位
Location

高差
Elevation difference

地形
Terrain

建筑
Architecture

净水池
Purification pool

路网
Road network

链接
Link

串联
In series

南岭 1932
| 从城市再生到共生的思辨

始建于伪满时期的南岭水厂，是当时工业文明的象征，它早已被城市的楼群遮蔽，这里本是一个纯粹的自然之地，却因为半个世纪的封闭管理，让其成为草木肆意生长的城市密林。当围界拆除，城市开放性的功能即将随之植入，这片水的痕迹是否蒸发成为"后现代都市"的模样？设计更倾向人作为自然系统的一个因子，以更加谨小慎微的态度去留住场地原有的时光印记、从而实现建筑、自然、城市和谐相融；于是这里有了枝繁叶茂的密林，穿梭在其中的廊桥连起了物理空间上的建筑、场所和森林，同时也连起了良性的生态环境及历史的记忆。这里是一处百年厂房的历史和今昔，全域化的生命和城市共同成长，铸造着一座千年的城市森林。

The Nanling Waterworks, which was built in the Puppet Manchurian period, has long been obscured by urban buildings.This is not a pure natural land. However, because of half a century of closed management, it has become urban forests with wild growth.When the perimeter is dismantled, open function of the city is about to be implanted. Will the traces of this water evaporate into the shape of a "modern city"?Designers prefer to take people as a factor of the natural system, to retain the original time stamp of the venue with a more cautious attitude, so as to achieve harmonious integration of architecture, nature and city; thus there is a dense forest. The shuttling gallery bridge connects the buildings, places and forests in the physical space and also connects the benign ecological environment and historical memories.

This is the history of a century-old factory building and today, the global life and the city grow together, casting a thousand-year-old urban forest.

池上的轻舞
旧水池上的亲水栈道

这是一个人与自然共生的场地。原来是一个被拿掉顶盖的钢筋混凝土盒子，蓄上了水，四面被植被围绕，随着时间推移，形成了一个小型生态圈。火炬树每年春天抽芽，秋天变成一抹红色；松树吸引了一只只松鼠、小鸟；水草芦苇在池边蔓延，野鸭、野鹅在水中嬉戏。如何让大家看到这片人工与自然和谐生存的场面，这是我们思考的出发点。调研时，设计师走过每一棵树，找到最佳的体验路径，最佳的观景角度。思考如何在最小的人工干预的基础上展现自然风貌，最后我们在场地内植入一条兼具交通与观景功能的栈桥链接了游客中心与树林、水面，让游客在游览体验的过程中寻找城市中的自然。

This is a site where man and nature co-exist.The site turns out to be a concrete box with a roof removed, filled with water and surrounded by vegetation, forming a small ecological circle over time.Torch trees sprout every spring and turn red in autumn; the pine tree attracts a squirrel, a bird; water grass reed in the edge of the pool spreads, wild ducks and wild geese play in the water.How to let everybody see this scene of harmonious co-existence of man-made and natural scene?There is our starting point.

When we investigated, we walked through every tree to find the best path to experience, the best viewing angle.Thinking about how to show the natural features on the basis of minimal manual intervention? Finally, we planted a trestle with traffic and view functions in the site. The trestle links the tourist center with the woods, the surface of the water, and allows visitors to look for nature in the city during the tour experience.

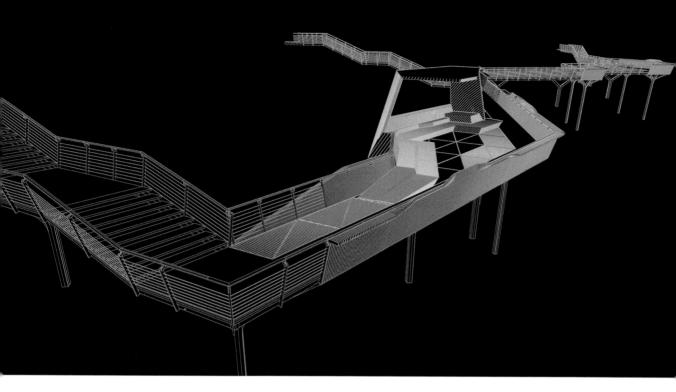

亲水栈道透视图
Perspectives view of the trestle

亲水栈道展开面

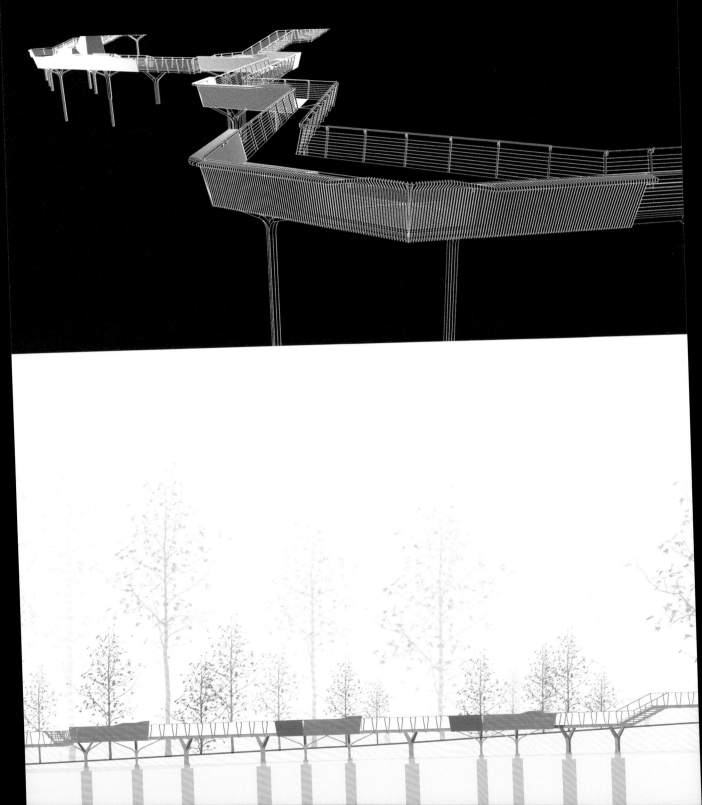

由露天沉淀池改造而成的生态休闲湿地,形成了建筑、植被、水体一体化环境
The ecological recreational wetland transformed from the open-air sedimentation tank forms an integrated environment of architecture, vegetation and water

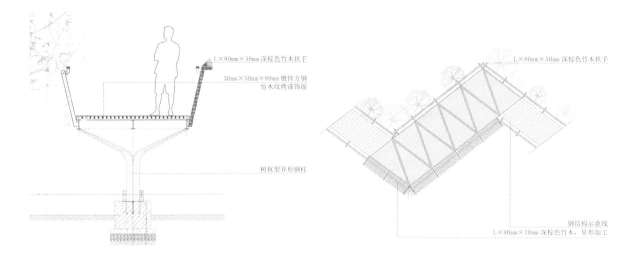

亲水栈道剖面图
Section drawing of the trestle

亲水栈道平面图
Plan of the trestle

亲水炬树共生的栈桥体系
Trestle system associated with torchbearers

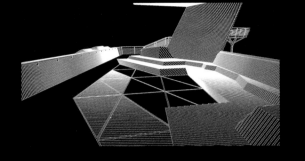
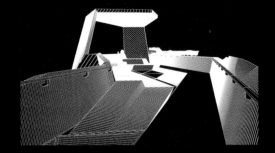

亲水栈道透视图
A perspective view of the trestle

夜晚灯光下的亲水栈道
A waterfront walkway under the lights at night

观景平台让游客可以驻足
The viewing platform allows visitors to stop

利用废弃的石材和钢管加固墙体结构
The wall structure is reinforced with waste stone and steel tubes

栈桥秋景
View of the trestle in autumn

鸟巢

| 建在树上的"家"

园区的自然精灵是栖息在林间的42种飞鸟，密林间、荒草丛是它们理想的巢穴，当你化作飞鸟，你梦想的巢穴在哪？设计师在寻觅最为适合的场地，畅想着与大树环抱、共同生长的巢穴，南露天沉淀池边上的两棵大杨树相伴相生，有着最好的望水视野。环绕式的楼梯形成了两层的望水平台，也将树与巢融为一体。结合安全防护的需求，增加了流动的曲线型格栅，进一步强化树上的鸟巢。

鸟巢更多地想传达融合共享的理念，园区内的人与物种都有着亲密的联系和归属感。

The natural elves in the park are 42 species of birds that inhabit the forest. The jungles and the grass are their ideal nests. When you turn into a bird, where is your dream nest? We are looking for the most suitable venue, thinking about the nest that surrounds and grows with the big trees. The two large poplar trees on the edge of the southern open-air sedimentation pool are accompanied by each other and have the best view of the water.The wraparound staircase forms a two-story water platform that also blends the tree with the nest.Combined with the demand of safety protection, the flow curve grille is increased, and the bird's nest in the tree is further strengthened.

The bird's nest wants to convey more about the concept of integration and sharing, and the people and species in the park have close connections and a sense of belonging.

两层的观景平台有着最佳的景观视野
The two-level viewing platform has the best view

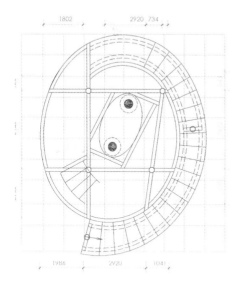

二层平台边界定位图
Location diagram of the boundary of the second floor platform

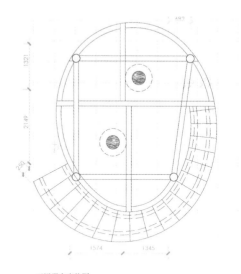

三层平台定位图
Three-layer platform positioning diagram

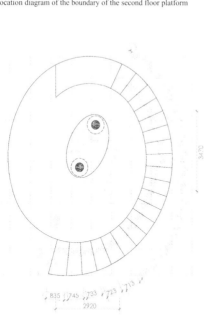

二层楼梯及树洞定位图
Second floor stair and tree hole location map

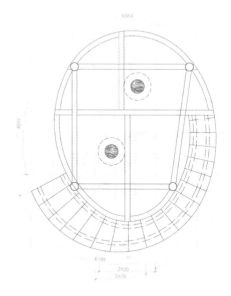

三层平台次龙骨平面图
Floor plan of secondary keel of three-storey platform

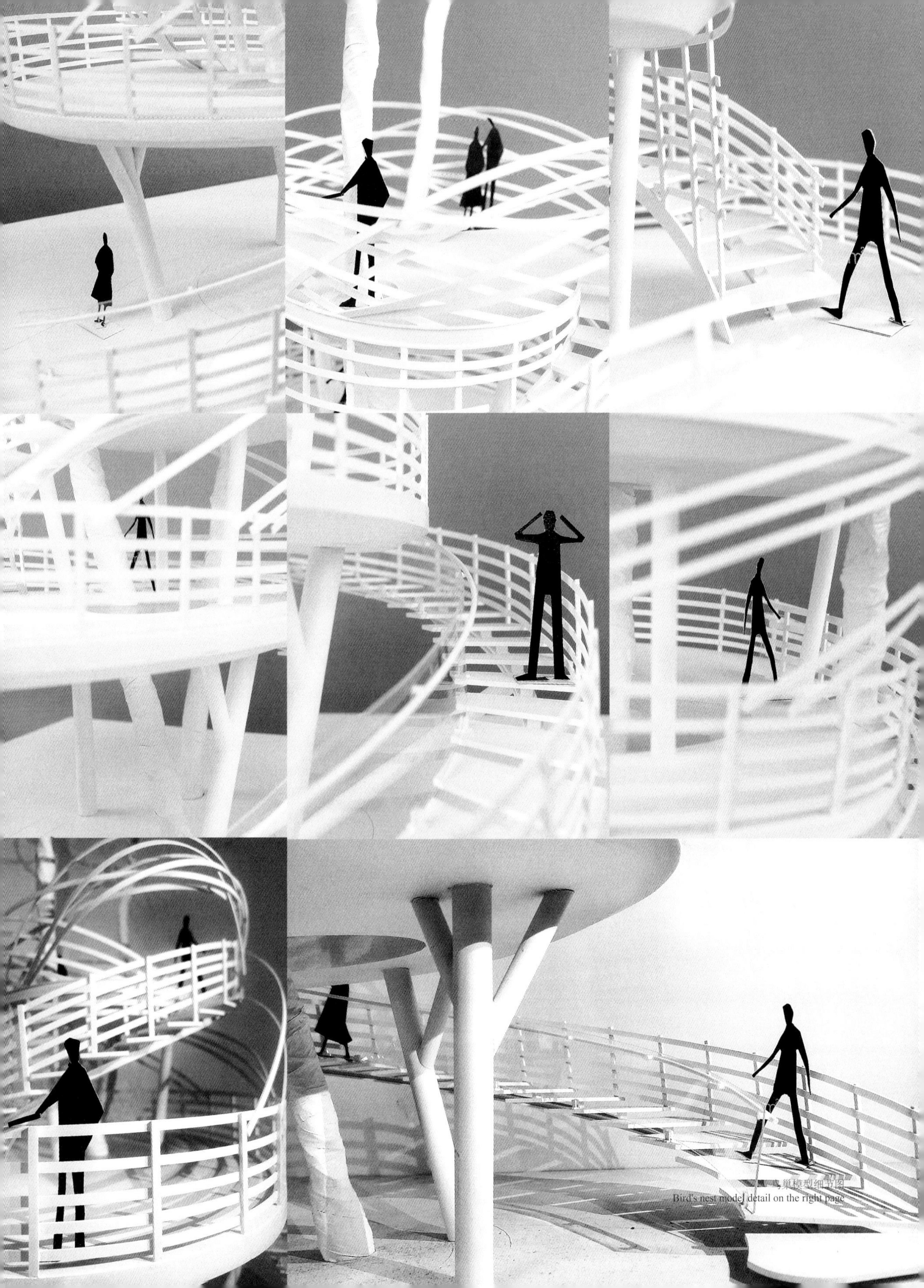

鸟巢模型细节图
Bird's nest model detail on the right page

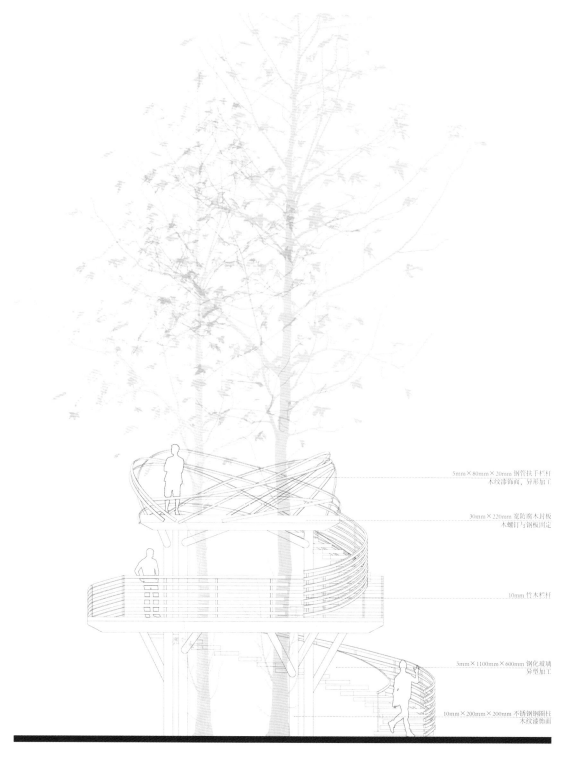

立面图
Elevation

两层的观景平台有着最佳的景观视野
The two-level viewing platform has the best view

生态连接线路径边的两棵大树
Two big trees on the path of the ecological chain

森林栈道
| 游走在林间

当生态链接线穿入郁闭的林带后，势必对生态造成巨大的冲击，那不如引一座桥，桥两侧起落的格栅主导着游走在桥上人的视线，这是设计师在数十次的现场调研中选定的故事主线，榫卯结构形式增加了桥工艺的细节之处。从桥上生长出来的树木，是设计者对自然的态度，落空的钢格栅栈道能够清晰地看见桥下郁郁葱葱的植被，那里是密林中动物的迁徙廊道。这里自然是主宰，行走是艺术。

When the ecological link line penetrates into the most closed forest belt, it is bound to cause a great impact on the ecology. It is better to build a bridge and grating rising and falling at both sides of the bridge dominates the line of sight of people walking on the bridge. This is the main plot of the selected by story the designer in dozens of on-site research. The mortise-tenon structure forms add details to the bridge process.The trees growing from the bridge are our attitude towards nature. On the overhead steel grille, we can clearly see the lush vegetation under the bridge, which is the migration corridor of animals in the jungle.It is dominated by nature here, and walking is art.

森林栈道 立面图
Elevation of the forest trestle

森林栈模型图
Model of the forest trestle

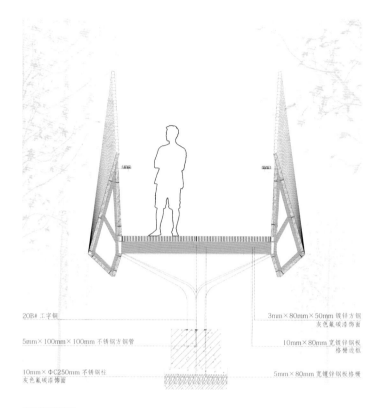

森林栈道剖面图
Forest plank road profile

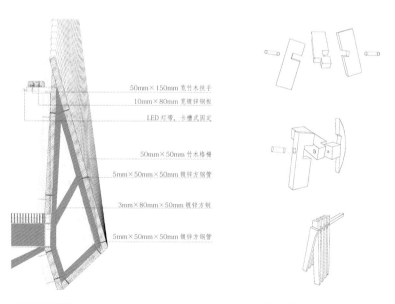

森林栈道格栅详图
Forest plank road grid detail

森林栈道格栅榫卯结构大样图
Large structural drawings of forest trestle grille falcons

密林里的架空栈桥为游客带来了多样的风景
The hanging trestle in the forest brings a variety of scenery for tourists

巨尺度森林巨人与森林栈桥形成呼应
The forest giant creates a dialogue with the forest trestle

森林纸鹤
隐藏在森林中的瞭望台

We can imagine folding a paper crane and making it fly into the forest on the hillside. It should be hidden in the forest and hard to be found. This is one of our ideas about the forest observatory, which is integrated with the forest. Based on this idea, we use a folded shape placed at the highest point of the hillside to map the surrounding environment through mirrored surface treatment. Fragmentation treatment is made as far as possible to ensure the flatness of the construction, reduce the special attention caused by defects. We formulate an interesting game rule for this folding paper crane. After walking up through the mirror of the narrow cave, we can see a group of buildings obscured by the forest. This is the oldest buildings of waterworks.

设想着折十只纸鹤飞向山坡上的森林，它该是隐在林间难以寻觅了。这是我们对林间瞭望台的一个畅想、和森林融为一体的。基于这样的想法，我们用了一个折叠的形体放置在山坡的最高点，通过镜面的表皮处理来映射周边的环境，尽量碎片化处理来保证施工的精致度，减少因为瑕疵产生的特别关注。我们给这个折叠的纸鹤制定了一个有趣的游戏规则，拾级而上穿过镜面的狭长山洞，一组绿意掩映的建筑群尽收眼底，这是水厂最老的建筑群落。

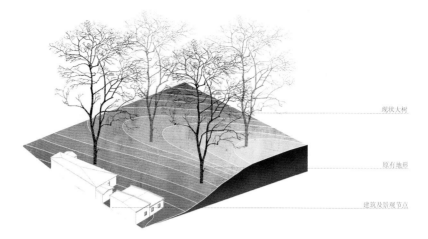

原有场地
The original site

选址
Site selection

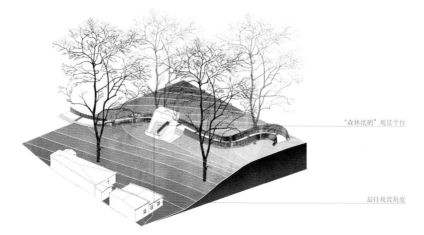

架设景观平台
Erection of landscape platform

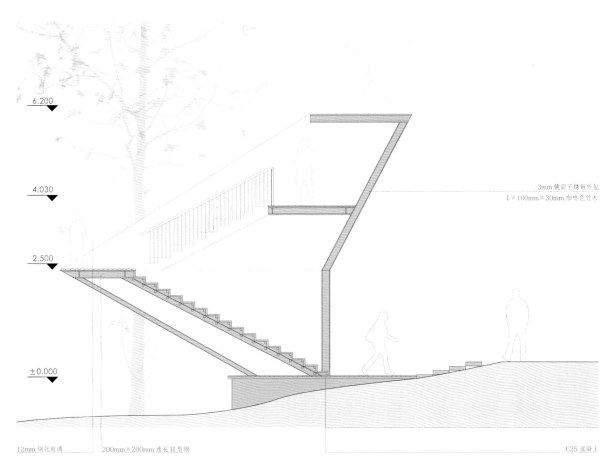

"森林纸鹤"剖面图
Section of "forest paper cranes"

碎片化的镜面表皮将森林纸鹤隐逸林间
The fragmented mirror skin hides "forest paper cranes" among the trees

漂浮的曲线
树梢上的森林桥

置身在这片垂直的柳树森林之中,生态链接线该是何种形态?设想着几条漂浮的曲线环绕在林间,让走在其中感受更加接近树叶、飞鸟和蓝天。那么这悬浮的曲线桥的装饰本身应该以更为简洁和轻盈的方式去呈现。柱子模拟树杈的形态希望能够尽量隐于森林。漫步间绕过树丛从地面走向树梢,沿着桥面和扶手设计师从照明上强化漂浮的感受,呈现了纯美的夜景效果。

When we are in this vertical willow forest, the ecological link line should be of certain form. We can envision several floating curves around the forest and have closer contact with the leaves, birds and blue sky while walking amid them.Then the decoration of this floating curve bridge itself should be presented in a more concise and light way.The column simulates the shape of the branches of trees in the hope of being hided as much as possible in the forest.Walking around the trees from the ground to the treetop, along the bridge deck and handrail, we enhance the feeling of floating from the lighting, showing a pure beautiful night view.

流动的栈道围合两棵树下空间，作为生态链接线上的休憩节点
The flowing plank road encloses the space under two trees and serves as a rest node on the ecological link line

林中游走的栈桥
Trestle in the woods

结合游览半径设置休憩空间
Setting the rest space according to the tour radius

贯穿于密林和公共空间的人行系统
A pedestrian system running through the forest and public spaces

林间花海
| 记忆中的那片白桦林

纯白的树干撑起的满眼金黄,是北方独有的白桦林带,山坡上的老厂房和这片树林成了水厂建筑与绿意融合最好的一个区域,基于现有情景,设计师更希望尽量少地融入强烈的人工形式感,一条木栈道、几件艺术化处理的老旧仪器是设计留下最重痕迹,接下来只需撒下一片种子,让大地染上一抹紫色的花海。

The pure white trunk is full of golden, and it is the northern unique birch forest belt. The old factory buildings on the hillside and this forest become the best area for integration of waterworks and greening. Based on the existing situation, we hope to integrate strong sense of artificial form as little as possible. A wooden walkway, a few old and artistically processed instruments are designed to leave the heaviest traces. Then we only need to sow seeds to make the land decorated by a purple sea of flowers.

成片的鼠尾草花海和白桦林
Groves of sage flowers and birch trees

废弃的枯木转化成了自然装置，与森林融为一体
The abandoned dead wood is transformed into a natural installation that blends into the forest

生态连接线串联现状散落的建筑和白桦林
The ecological chain connects the scattered buildings and birch forests

旧石料转化成公园的铺装材料
The old stone was converted into pavement for the park

用废旧净水设施改造的艺术装置被散布在白桦林、一片鼠尾草花海和金黄的落叶之间构造出一年四季的风景。
Art installations that have been converted from old water treatment facilities are scattered among a grove of birch trees, a sea of sage flowers and golden leaves to create a seasonal scene.

艺术广场
Art Square

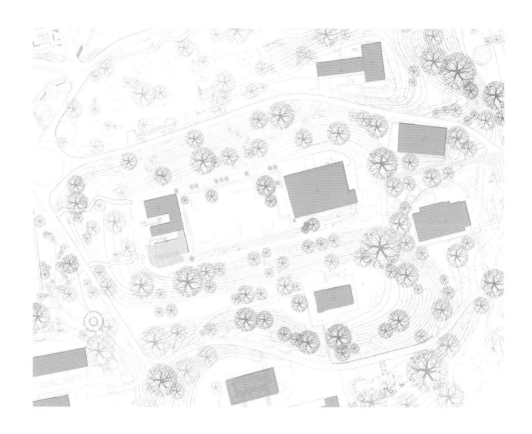

遗迹的新生
| 水池上的艺术空间

老水厂最为核心的区域，以净水池为核心的建筑组团，设计师更希望它可以承载水厂更加多元类型的活动。于是设计师清掉池盖上的杂草，作为一片综合活动的草坪进行呈现。拆开池子的柱子放置在场地上，与地下的空间形成一个文化的呼应。彩色的雕塑和白色的艺术廊架丰富了空间的现代艺术气息，老梁、爬山虎、草坪、现代艺术品被围在一个山谷里，这里既是城市活力活动的空间，也是森林中难得的阳光浴场。

For the core area of the old waterworks, with the clear water pool as the core of the building group, we wish it to be able to carry more diverse types of activities in the waterworks. So we cleared the weeds on the pool cover and presented them as a comprehensive activity lawn. We disconnected the pillars of the pool and place them on the ground, forming a cultural echo with the underground space.

The colorful sculptures and white art gallery enrich the modern art of the space. The old beams, ivy, lawns and modern art are surrounded by a valley. It is both a space for urban vitality and a rare sun bath in the forest.

林间的子时征艺术广场：承载丰富市民活动
Multi-functional art square in the maple carries a wealth of civic activities

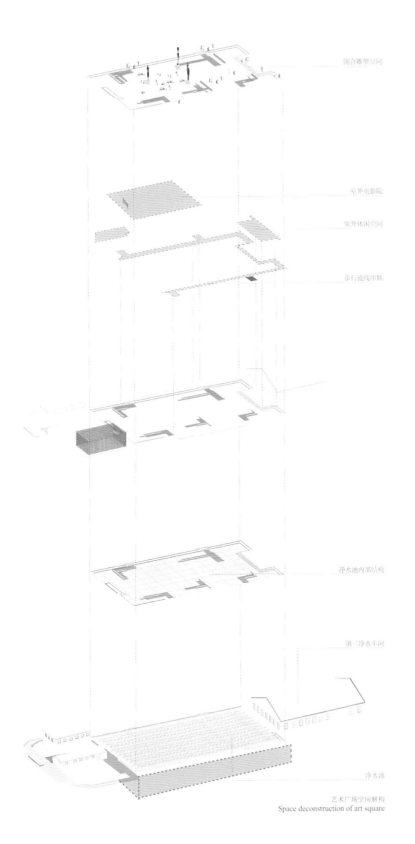

围合雕塑空间

室外电影院

室外休闲空间

步行流线串联

净水池内部结构

第二净水车间

净水池

艺术广场空间解构
Space deconstruction of art square

建成前
Built before

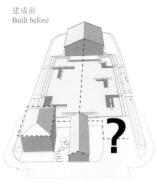

建成后
After establishment

利用沉淀池顶部建造的多功能艺术广场,植入当代艺术装置
A multifunctional art plaza built from the top of the sedimentation tank is implanted with contemporary art installations

艺术装置及工业遗迹空间
Art installation and industrial relic space

白色艺术装置映射着周围环境
The white art installation reflects the surroundings

树林中的艺术剧场:具有多元化功能的公共活动空间
In the art theater, which has the function of diversified public space

站在一片绿意盎然，森林般树木掩映的空地中，面对一座40年前的房屋，人们会产生一种年代的错觉。如果这是一个梦境，当一片森林置身于城市时，人们会感觉它很无助甚至恐惧，但城市置身于一片森林中，人们会感觉它拥有了很强的生命力。

设想将一个线性的建筑体置入一个原始基底的平面上，通过对功能场景的重新定位，形成生动的生活单元，而这些单元组成城市生活盒子来展现未来城市人群的活动可能性。生活盒子由不同的活动组成，局部在材料上使用镜面不锈钢材质，反射周边植物环境，产生一种时代与环境的新型语言，形成一个互动的有机载体。生活盒子的呈现将大大增强人与人，人与自然之间的新的关系，减少彼此之间的距离，让更多的人在自然中体会城市，感受幸福，塑造归属。

Standing in a wooded clearing, facing a house built 40 years ago, you get a sense of age. If this is a dream, when a forest in the city downtown, you will feel it is very helpless and even fear, but when the city in a forest, in the city you will feel it has a strong vitality. We imagined to place a linear building body on the plane of an original base, and form vivid living units through repositioning the functional scenes, and these units formed urban living boxes to show the activity possibilities of future urban population. The living box is made up of different activities. Specular stainless steel materials are used locally to reflect the surrounding plant environment, creating a new language of The Times and the environment, and forming an interactive organic carrier. The presentation of the life box will greatly enhance the new relationship between people, people and nature, reduce the distance between each other, let more people experience the city in nature, feel happiness, and shape attribution.

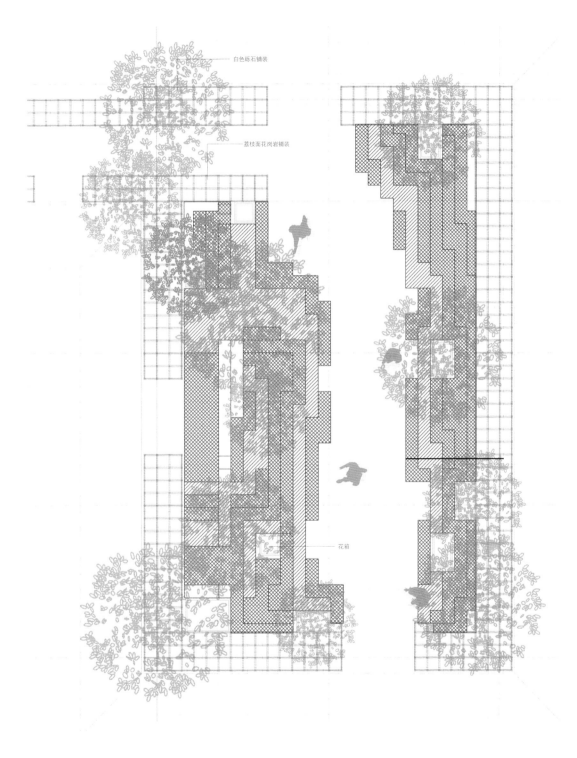

白色艺术装置平面图
Plan of the white art installation

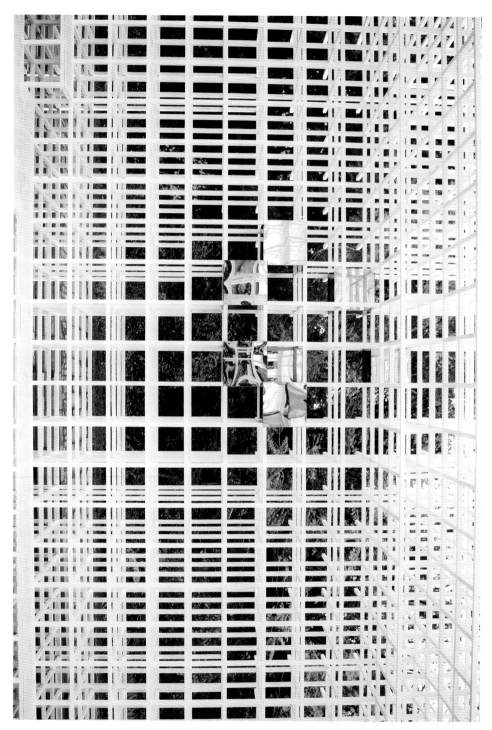

白色艺术装置细节
Detail of the white art installation

艺术装置及工业遗迹空间
Art installation and industrial relic space

A multifunctional lawn built from the top of the sedimentation tank is inserted into the contemporary art installation

遗迹的新生
| 旧水池里的生长空间

十个带顶盖的沉淀池是园区最具特色的工业遗迹,沉寂在地下的空间有着太多的未知性,作为唯一一个被打开顶盖的池子,它有太多的东西需要承载和表达。设计师掀开池子的过程充满了各种不确定性和惊喜。满是水锈的墙体至今保留着斑驳的肌理,水池底原有的柱子作为场地的遗迹大部分保留下来,比如:原有两米高的通风道被作为景观通道,通道下的墙体将水池分成两个空间,一侧结合雨水花园和趣味净水设施的体验区;另一侧是柱阵与喷水雕塑形成的艺术空间。种在池子里的一栋小建筑增强了下沉空间的活力,钢格栅的立体交通让上下两层空间更加连通。旧的池壁、钢板、竹木、玻璃交织杂糅互为支撑却不突兀、苏醒的池子将北侧仪式感的老建筑和南侧灵动的小建筑糅合在一起,共同诉说这水与城市、水与活力、水与艺术的和谐之美。

The ten covered sedimentation pools are the most characteristic industrial relics in the park. The space buried in the ground has too much unknownness. As the only pool that has been opened, it has too many things to carry and express.The designer's process of unlocking the pool is filled with uncertainty and surprises. The rusty walls still have mottled texture. Most of the original pillars at the bottom of the pool are preserved as the remains of the site. The original two-meter-high ventilation duct is preserved as the landscape passage. The wall under the duct divides the pool into two spaces. One side is experience area combining rainwater garden with interesting water purification facilities and the other side is the artistic space formed by column array and water spraying sculpture.A small building planted in the pool enhances the energy of the sinking space, and the three-dimensional traffic of the steel grille makes the upper and lower layers more connected.The old pool wall, steel plate, bamboo, and glass intertwine with each other but are not awkward, and the awakened pool integrates the old building on the north side and the small smart building on the south side to jointly present the beauty of harmony between the water and the city, water and vitality, water and art.

下沉雨水花园剖面图
Section of sunken rainwater garden

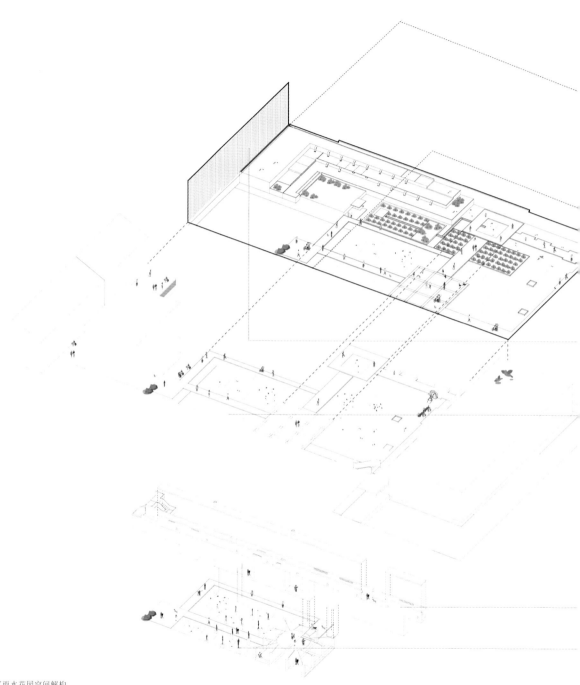

下沉雨水花园空间解构
Spatial deconstruction of sunken rainwater garden

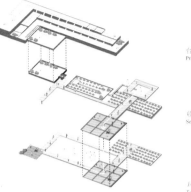

台地式 一级净化过滤
Primary purification filtration

收集式 二级净化过滤
Secondary purification filtration

环地式 三级净化过滤
Tertiary purification filtration

水与净化
依据地形引导雨水径流，设计台地式
雨水花园景观层层净化

水与生活
液态融入洁水池，激活水池

水与艺术
雕塑与内部结构的融合与抬升的艺术装置为人群
提供了浓厚的艺术氛围

雕塑

由封闭水厂沉淀池改造成的下沉公共空间
A sunken public space transformed from a closed water plant sedimentation tank

利用沉淀池高差形成的雨水花园
The rainwater garden formed by using height difference of sedimentation tank

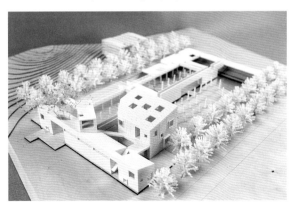

下沉雨水花园模型透视图
Details of the sunken rainwater garden mode

水净化与水利用的空间对话
Spatial dialogue between water purification and water utilization

保留原有的柱网结构
Retain the original column network structure

由封闭沉淀池改造成的下沉公共空间
A sunken public space transformed from a closed sedimentation tank

下沉公共空间及历史建筑的一体化保护利用
Integrated conservation and utilization of sunken public space and historic buildings

充分保持原有的历史痕迹,并赋予功能化处理
Fully preserving the original traces of history and giving functionalized processing

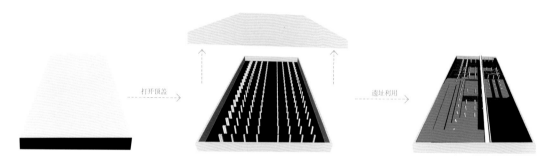

原有封闭沉淀池打开顶盖
Open the top cover of the original closed sedimentation tank

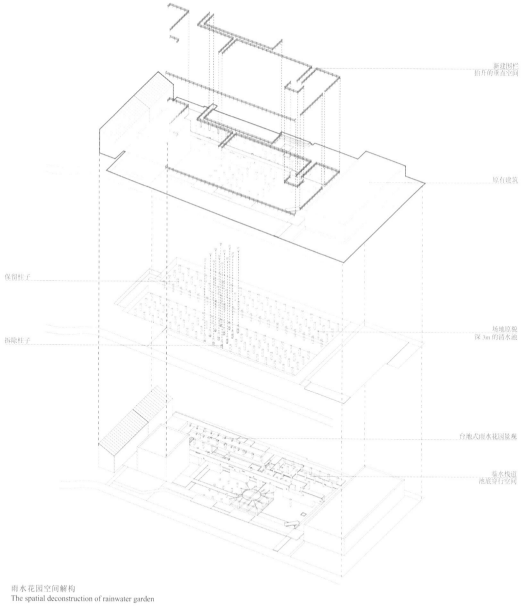

雨水花园空间解构
The spatial deconstruction of rainwater garden

Site inspection of sunken rainwater garden

旧有通风廊道改造的通行空间
The passage space transformed by the old ventilation corridor

整体鸟瞰
A bird's eye view of whole

水文化博物馆
Water Culture Museum

记忆空间
| 水的新奇科普之旅

南岭，1932 是从这组满是爬藤的建筑区发源而来的，三个日伪时期的池子上是古董一样的净水设施，水文化博物馆的功能植入是希望赋予老建筑以新的生命和活力。希望通过一种方式将被建筑包围的三个池体连一起与建筑形成一个历史感，以整体呈现，原有一个池体上面长满了野草的景象被保留下来，结合动物雕塑彰显生态的特征；最老池体清掉上盖上的覆土，把古老的结构呈现出来；靠近道路的池体上盖融入互动性的戏水设施，作为博物馆入口最具趣味的水科普乐园。

Nanling 1932 originates from the building zone filled with vine and ancient water purification facilities are on the three pools in the puppet Japanese period. Function implantation of the Water Culture Museum is to give new life and vitality to the old buildings.The landscape aims to link the three pools surrounded by the building and form a historical entirety. The original scene of a pool full of weeds is preserved to highlight the ecological characteristics combined with animal sculpture; after clearing the cover soil on the upper cover the oldest pool body, the ancient structure is presented; the upper cover of the pool near the road incorporates interactive water-playing facilities, which is the most interesting water science park at the entrance of the museum.

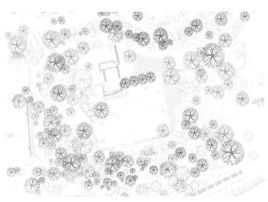

保留现状建筑
Preservation of existing buildings

新增装置
New device

水文化重塑
Water culture remolding

道路体系
Road system

场地生成
Field generat

水文化博物馆空间解构
Spatial deconstruction of water culture museum

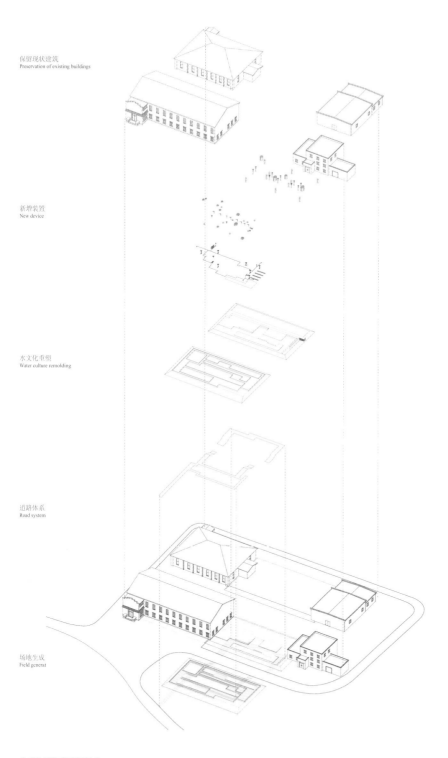

在古老的池体上，游人与污水装置的互动
Visitors interact with water purification devices in the ancient pool

被保留的工业遗迹
A preserved industrial relic

水文化博物馆鸟瞰
A bird's eye view of the water culture museum

旧有净水设备的新生
A rebirth of old water purification equipment.

对原有建筑的保护和修缮
Protection and renovation of existing buildings

儿童活动场地
Children's Activity Area

密林中的活动场地
| 都是林间的野孩子

This is a venue with theme for children.We unfold this design from the memory of the venue, and at the same time think about the connection between the venue and the children activity space. Where is the connection?First of all, this site is close to the eastern residential area, carrying the interaction between urban life and the site "artificial nature". Secondly, the original site of the site is the playground with low utilization rate in the waterworks, and it is also the only hard site that is not covered by vegetation in the waterworks.All this means that we can implant a functional site without destroying the ecology of the site.

If this is just a simple children's paradise, except the location of the site, where is the memory of the waterworks? There are still a lot of industrial relics in the waterworks, which provide us with endless design materials.A water valve that can be turned, a repainted can, and a large water pipe that can accommodate people can be their paradise.What we need to do is to recollect the materials, reorganize them, and leave the rest to the children who will open a new world of their own.

这是一个关于儿童主题的场地。设计师从场地记忆上展开这个设计，并且同时思考场地和儿童活动间到底有什么联系，联系在哪里？首先，这块场地是靠近东面住宅区的场地，承载着城市生活与场地"人工自然"的互动；其次，场地的原身是水厂里利用率很低的运动场，也是水厂里唯一的不被植被覆盖的硬质场地。这些都注定我们不需要破坏场地的生态，就可以植入一个功能性的场地。如果这只是一个单纯的儿童乐园，除去场地的位置因素，那水厂的记忆又在哪里？水厂厂房内还保留了大量工业遗迹，这为设计师提供了无尽的设计素材。一个还可以转动的水阀，一个重新上色的罐子，一根可以钻进人的大型水管，这些都可以成为他们的乐园。设计师需要做的，只是把这些素材重新收集，重新组织，剩下的就交给孩子们，他们会自己打开一个新世界。

A children's playground under a thick forest

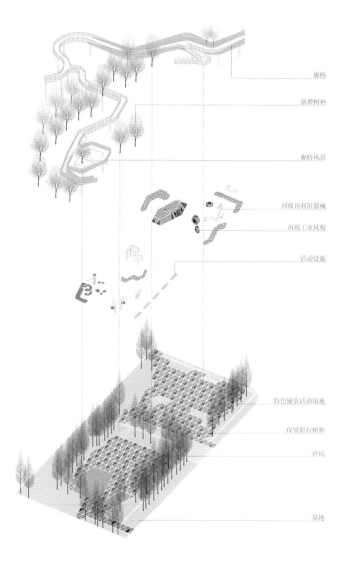

廊桥
新增树种
廊桥风景
回收再利用器械
再现工业风貌
活动设施
特色铺装活动场地
保留原有树种
沙坑
基地

儿童活动场地空间解构
Spatial analysis of children's activity space

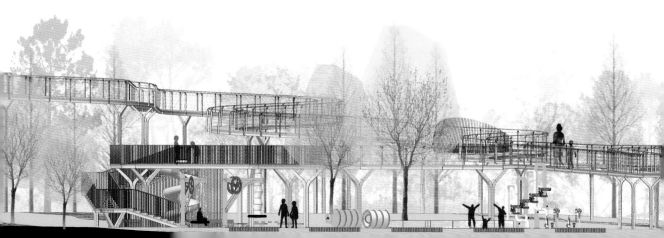

旧的净水设施被改造成为儿童的互动游戏设施
The old water purification facility has been transformed into an interactive play facility for children

儿童活动场地立面图 下
Elevation of children's activity area

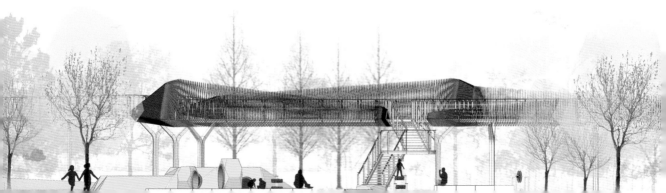

改造后的净水厂成为市民可参与、感知的活力空间
The transformed water treatment plant has become an active space for the public to participate and feel

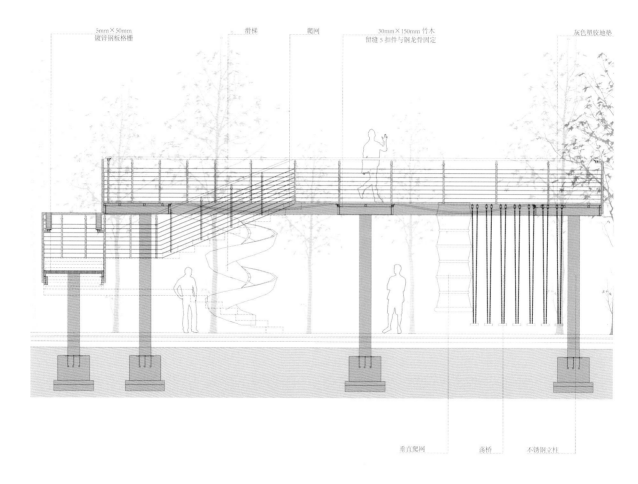

儿童活动场地立面图
Elevation of children's activity area

与空中栈桥连接的儿童设施
A children's facility connected to an aerial trestle

圆形沉淀池
Circular Sedimentation Tank

圆舞曲
| 大地的畅想

水厂内的很多功能场地是在地下的,再加上场地自身多年的自然生长,想要完全展现水厂的功能场地是十分艰难的。这是一场浩大的工程,工程中必然会破坏场地良好的生态,无疑会成为灾难。在这个场地的设计中,我们试图在形态上再现地下圆形清水池,让它的形态成为我们设计的母题。同时融入大地艺术的设计思路,让场地本身成为一个公共艺术装置。

Many of the functional sites in the waterworks are underground, and with the natural growth of the site itself for many years, it is very difficult to fully display the functional site of the waterworks.This is a huge project. The project will inevitably destroy good ecology in the site and there is no doubt that this will become a disaster.In the design of this site, we try to reproduce the shape of the underground circular clear water pool, let the shape become our design motif.At the same time, we incorporate the design thought of the earth art, so that the site itself can become a public art installation.

圆形元素赋予丰富的空间体验
The round elements give a rich spatial experience

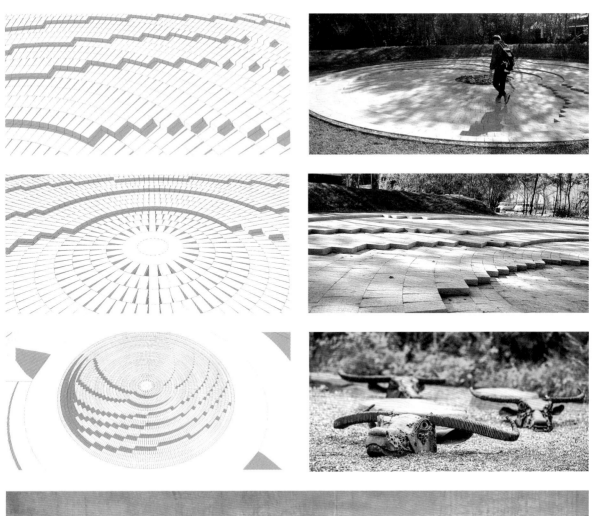

圆形水剧场的元素演变
The evolution of the elements of the water amphitheater

133

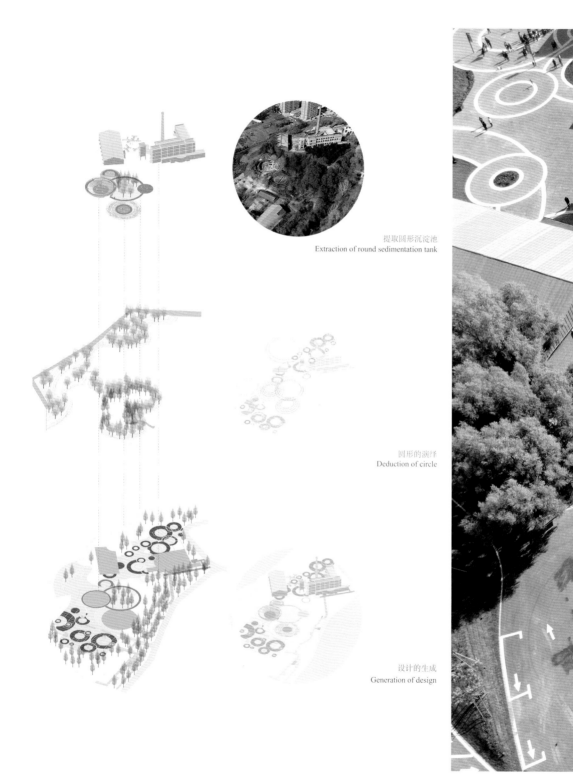

提取圆形沉淀池
Extraction of round sedimentation tank

圆形的演绎
Deduction of circle

设计的生成
Generation of design

圆形沉淀池空间解构
Spatial deconstruction of circular sedimentation tank

利用沉淀池高差形成的雨洪花园
The rain flood garden is formed by using height difference of sedimentation tank

圆形沉淀池鸟瞰
A bird's eye Bird view of round sedimentation tank

古树名木区
Area Of Ancient And Famous Trees

百年名木区
| 树下的冥想

实地踩场来到这个有着历史遗留下来的老树山丁子，高杆垂枝，树冠茂密，冠伞很大，宛自天开，白色花朵遍布如雪，每一个姿态都是水厂历史的印记，设计师绕树走了一圈，抬头看这光影从叶间洒下，很想躺在树下沐浴阳光。是的，应该在它的下面设计一个既可以发呆可以玩的空间，以自然为本的哲学，遵循自然生长的时间与速度，创造等待花开，等待草原的滋长，等待红叶，等待日落的宁静景观，唤起冥想体验。环绕老树，座椅以拟生的手法模拟垛子的形态，辅以木质生态材料。靠背尺寸高低变化，高可作为两排座椅，低可作为扶手。座椅的宽度适宜两个人横躺，舒适的感受自然的共生。

On the spot, I came to this old tree stalk with history. There were high trunks, drooping shoot, dense and large crowns stretching naturally. There were many white flowers and every gesture is the imprint of the life of the historical waterworks, I walked around the tree, looked up and saw the light and shadow from the leaves. I wanted to lie under the tree to soak up the sun. Yes, I should design a space underneath it for dazing and playing. According to nature-based philosophy, I follow the time and speed of natural growth, create tranquil scene for waiting for flower blossom, waiting for the growth of the grassland, waiting for the red leaves, waiting for the sunset, thus evoking a meditation experience. Simulate the shape of the buttress with a lifelike method around old tree and seat, supplemented by wooden ecological materials. Backrest size changes from high to low. If it is adjusted to be high, it can be used as a two-row seat. If low, it can be used as an armrest. Seat width is suitable for two people to lie side-by-side and feel natural symbiosis comfortably.

座椅立面
Elevation of the seat

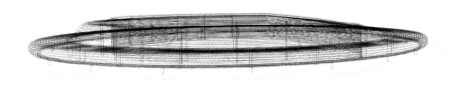

座椅空间解构
Spatial structures of the seat

座椅断面
Section of the seat

百年的山丁子树见证园区 80 年的岁月变迁，起伏变化的弧形树池呼应古树舒展的姿态，为市民提供了感受历史与自然对话的社交场所。

The century malus baccata borkh tree witness the changes of the park in the past 80 years.The undulating arc tree pool echoes the spreading posture of ancient trees,provideing a social place for citizens to feel the dialogue between history and nature.

树梢上的漫步道 | 西安白鹿原西坡树梢漫步栈道
Rambling Road At The Treetop
Rambling Plank Road At The Treetop Of Western Slopes Of Bailuyuan XiAn

地点 | 陕西省西安市
规模 | 70万平方米
业主 | 西安浐灞管委会

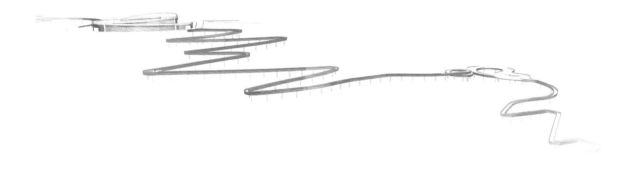

树梢上的漫步道
织路——织的是风景，织的是生活

"白鹿原上的项目"，第一次听到这个项目名称时，在内心打了一个大大的问号，带着疑问初次走进了陈忠实老师笔下那有血有肉，有情有意的黄土地。眼前的场景深深吸引了设计师。场景是震撼的，但也是忧伤的，震撼的是那千百年来的白鹿原的台原地貌，忧伤的是眼前的一切不是我书中读到的麦田圣皇，熙熙攘攘的盛景，更像是一块干瘪的海绵等待一场春雨的滋润。设计的出发点也顺着这样的场景走去，希望让这片土地重新焕发出那浓郁的历史特征，重新承载人们对白鹿原的期许及历史记忆的形态。同时希望打破狭义的景观设计范畴，通过一种复合的设计语言，将文脉空间环境通过再造再生。试图通过一个文脉的栈道，将历史的文化资源，人文资源，生态环境相连，充分释放。去诠释一片土地时历史文脉修复，区域形象文化记忆，区域经济的共融共生的概念。

When I first heard the name of "Project in Bailuyuan", I made a big question mark in my heart and entered the flesh-and-blood yellow land written by Chen Zhongshi for the first time.I was very taken by what I saw.The scene is shocking, but also sad. The reason why I am shocking is because of platform terrace of Bailuyuan for thousands of years. The reason why I am sad is because everything is not wheat field and the bustling scene read from the book. It is more like a dried-up sponge waiting for moistening of a spring rain.

The starting point of the design also follows such a scene. I hope that the land can re-revitalize the rich historical features, re-bearing the people's expectations and shape of historical memory.At the same time, I hope to break the narrow scope of landscape design and regenerate the context space by means of a compound design language. I'm trying to connect historical cultural resources, human resources, and ecological environment through a path of the context to realize full release.I also want to interpret the concept for restoration of the historical context of a piece of land, cultural memory of the regional image, regional economic integration and symbiosis.

白鹿原平面图
White deer plain plan

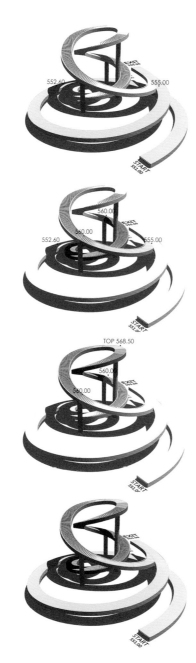

白鹿驿站结构受力点分析
Analysis of structural stress point of white deer station

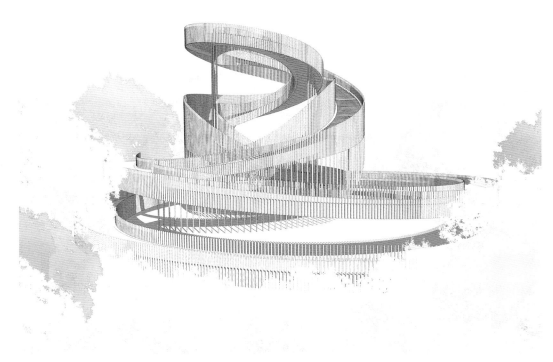

白鹿驿站透视图
Perspective of white deer station

白鹿驿站透视图
Perspective of white deer station

鹿原驿站平面图
Plan of white deer station

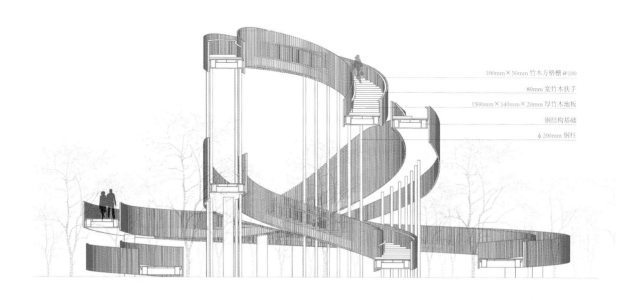

100mm×50mm 竹木方格栅@100
80mm 宽竹木扶手
1500mm×140mm×20mm 厚竹木地板
钢结构基础
φ200mm 钢柱

鹿原驿站剖面图
Profile of luyuan post station

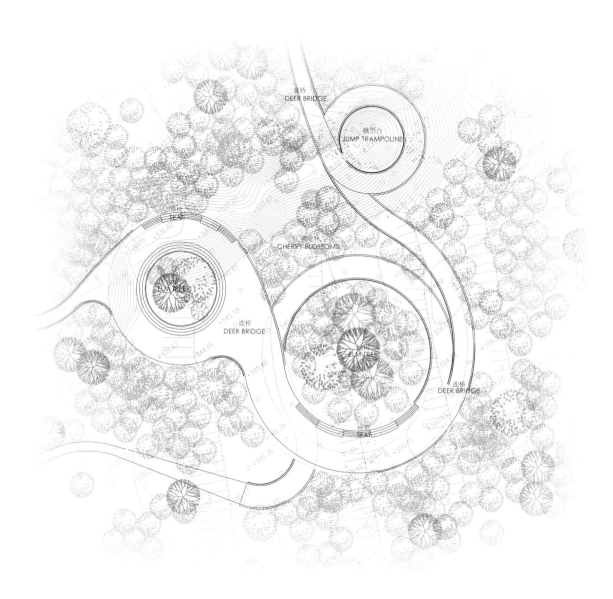

鹿原揽樱平面图
Deer original pull sakura plan

鹿原驿站模型
Model of luyuan post

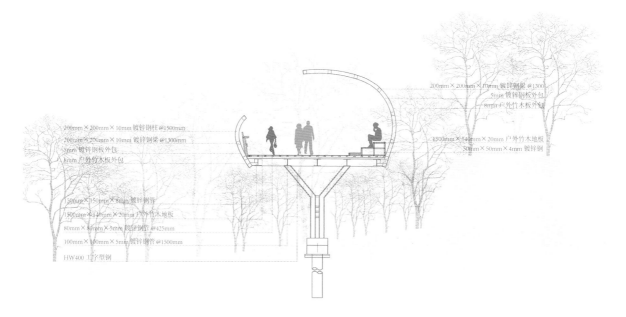

鹿原揽樱剖面图
Deer yuan lala sakura profile

鹿原揽樱模型
The deer pulled the cherry model

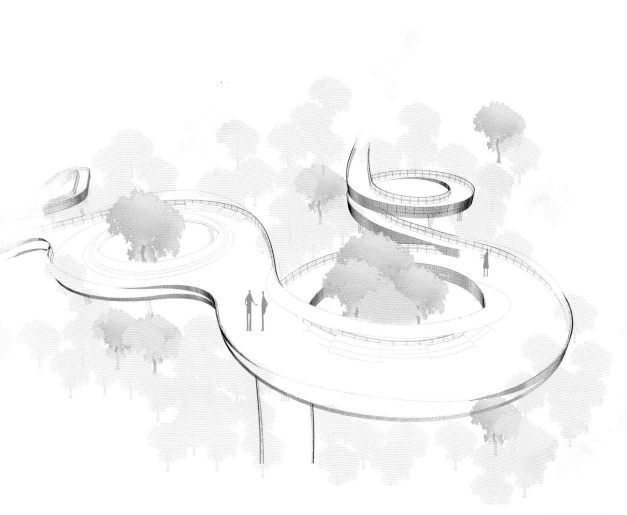

鹿原驿站透视图
Perspectives of luyuan post

游客服务中心
Visitor center

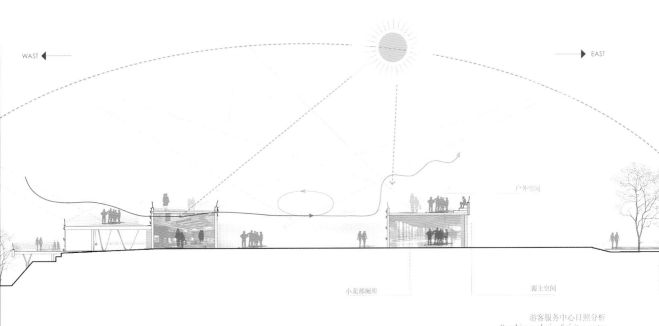

游客服务中心日照分析
Sunshine analysis of visitor center

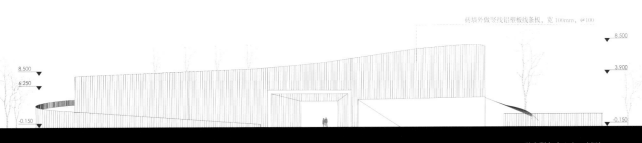

游客服务中心入口剖面
Section of entrance of visitor center

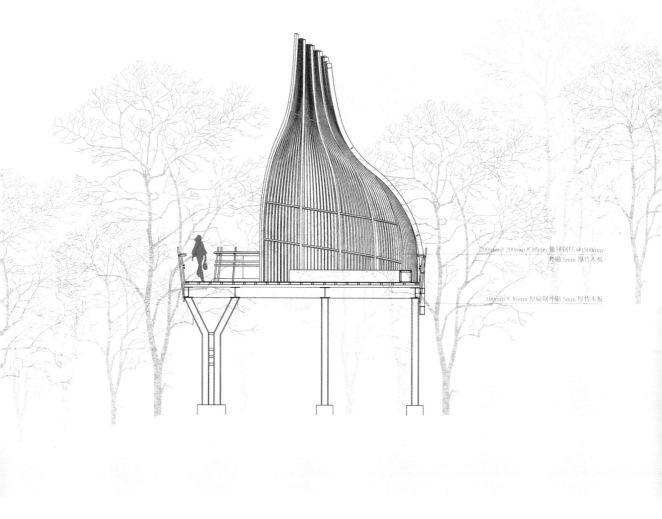

鹿亭剖面
Section of deer-pavilion

鹿亭效果图
Perspectives of deer-pavilion

鹿桥风雨廊剖面
Section of deer-bridge corridor

鹿桥风雨廊效果图
Perspectives of deer-bridge corridor

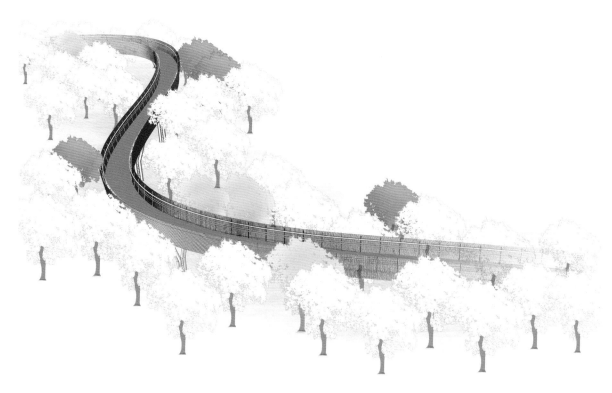

鹿桥效果图
Perspectives of deer-bridge

鹿桥标准段模型
Model of standard segment of deer-bridge

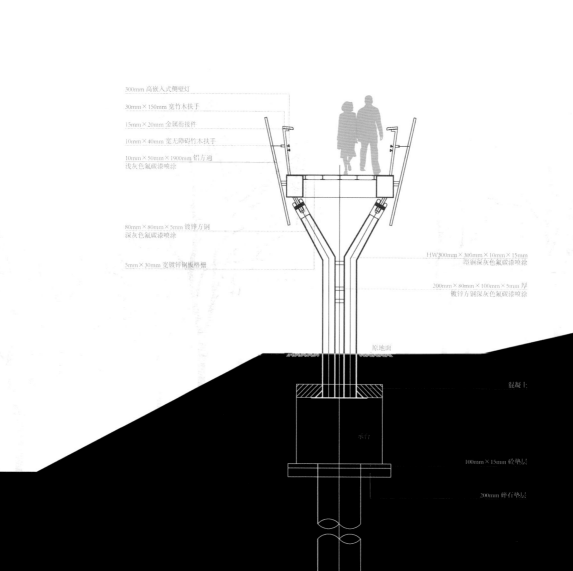

鸟瞰效果图
Bird's eye view effect

城市的微更新 | 上海田林路改造
Small Update Of The City
ShangHaiTianlin Road Renovation

地点 | 上海市
规模 | 1.1 万平方米
业主 | 上海市徐汇区田林路街道办事处

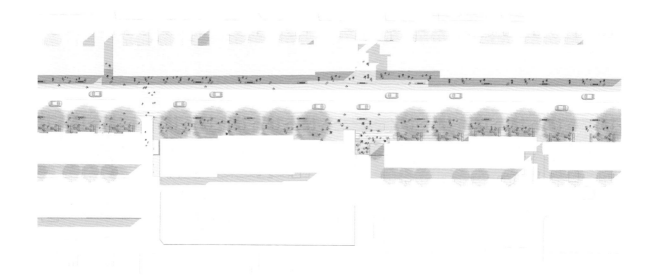

在共生街区中寻找城市的活力
从上海的魅力在梧桐树下的街道中，在每片落叶的赞歌

我们生活的环境，每天都有生命不断更迭的自然现象。城市也一样，它会有属于自己的生命周期，一个城市由若干的建筑和街道组成。街道也往往是构成城市中最重要的风景之一，同时也是"人生有志，先争朝夕"的载体。我们设计街道，不能只关注设计形式本身，应该从城市进化的逻辑去思考街道（区）的演化。

街道其实就像一块巨大的海绵，可以自然地接纳陌生而又新鲜的"水份"，同时与街道、建筑、商业、环境等形成有机的载体，让生活与场景共融，为使之成为一种可持续的共生状态。这种持续发生的进行和转变对居民生活多样及品质的提升提供了肥沃的土壤。街道（区）共生的概念也由此而生，它不仅代表了我们设计的终点，也代表了我们的初衷。场地调研时驻足观看，街道表达出的时快时慢的生命运动，它不仅属于片刻与永恒，更是一种人与环境的亲密体验。街道细分有很多相对独立的单元，每一个独立的单元可视为一种独立的体验，是人与环境、空间的一次对话，这些对话都从太阳，光线，树木，建筑，光影中体现，同时也是这些因素将街道空间一一激活。有人说梧桐树下才是上海。那隐藏在梧桐树后面的活色生香又是什么呢？

The environment in which we live has a natural phenomenon of changing lives every day.The same is true for cities, which have their own life cycle. A city consists of several buildings and streets.The streets also often constitute one of the most important scenery in the city, and they are also the carriers of "life is full of aspirations. We should fight for life".While designing streets, we should not only focus on the design form itself. We should think about the evolution of streets (blocks) from the logic of urban evolution.
In fact, the street is like a huge sponge. It can naturally accept strange and fresh things, and forms a systematic carrier with the street, architecture, commerce, environment and others, so that life and the scene can be integrated to become a sustainable symbiotic state.This ongoing process and transformation provides a fertile ground for the diverse living and quality of the residents.The concept of street (block) symbiosis has also emerged from this. It not only represents the end of our design, but also represents our original intention.At the time of research on the site, I stopped to watch, and the street expressed the fast and slow life movement. It not only belonged to the moment and eternity, but was also an intimate experience of people and the environment.Street segmentation has a number of relatively independent units and each independent unit can be regarded as a separate experience. It is a dialogue between people and the environment, space. These dialogues are reflected from the sun, light, trees, architecture, light and shadow. It is also these factors that activate the street space one by one.Some people say that Shanghai is only under the sycamore tree.What is the fragrance that is hidden behind the phoenix tree?

No.03

上海田林路改造

LANDSCAPE DESIGN ABOUT TIANLIN ROAD RENOVATION

2018
SHANG HAI

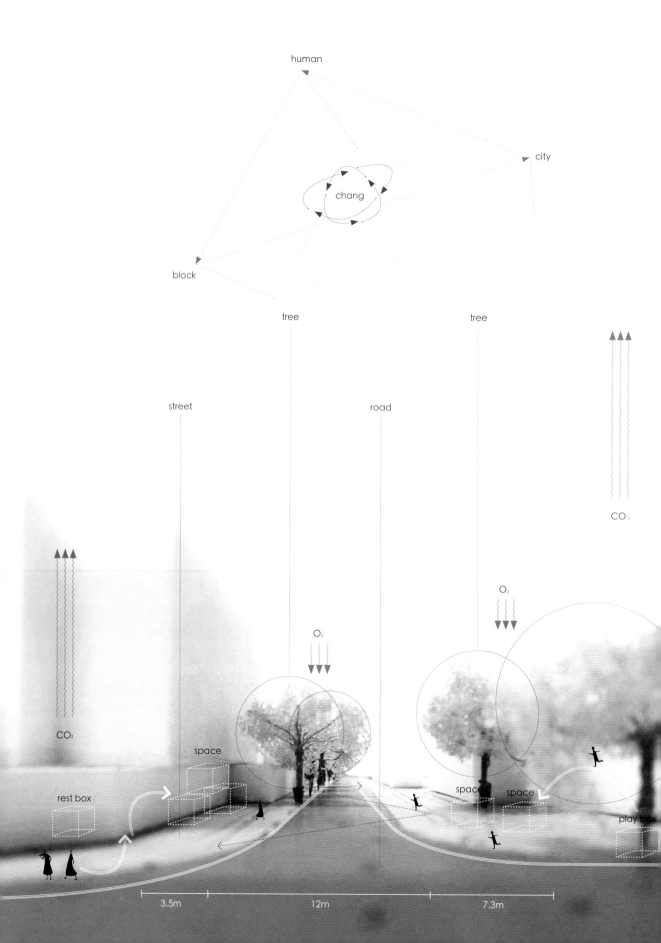

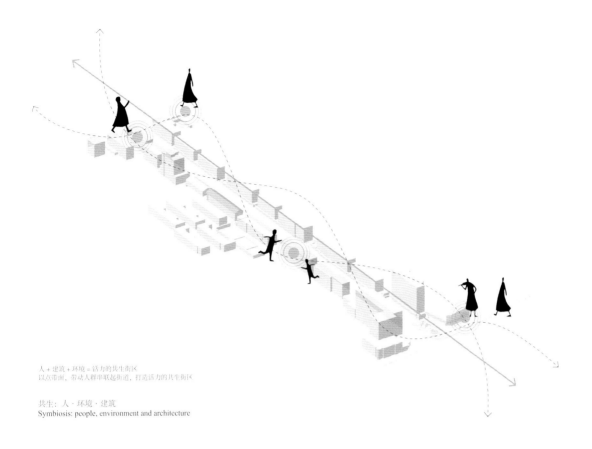

人 + 建筑 + 环境 = 活力的共生街区
以点带面，带动人群串联起街道，打造活力的共生街区

共生：人·环境·建筑
Symbiosis: people, environment and architecture

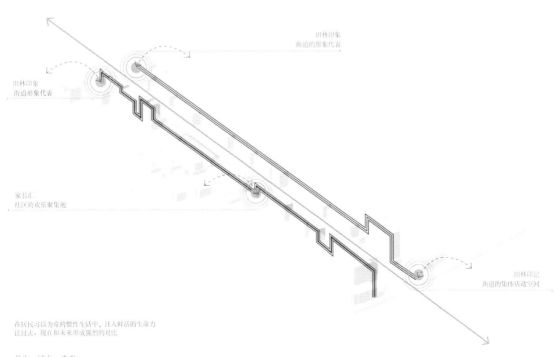

在居民习以为常的惰性生活中，注入鲜活的生命力
让过去、现在和未来形成强烈的对比

共生：过去·未来
Symbiosis: past and future

人与建筑
People and architecture space

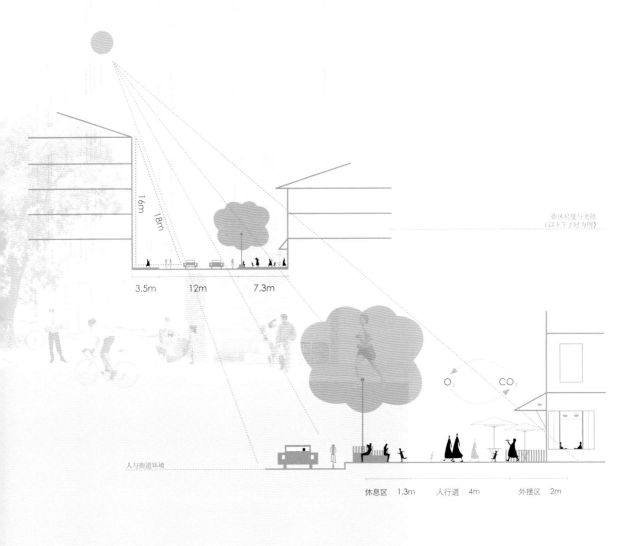

人·环境·建筑的关系
Relationship between people, environment and architecture

街道现状 人·建筑与环境·三者相脱离
The current situation of streets, people, buildings and environment which is separated from each other

生活和场景共融
Life and scene blending

街道其实就像一块巨大的海绵,可以自然地接纳陌生而又新鲜的事物,同时与街道、建筑、商业、环境等形成有机的载体,让生活与场景共融,使之成为一种可持续的共生状态。这种持续发生的进行和转变为对居民生活多样及品质的提升提供了肥沃的土壤。

街道改造效果图
Perspectives of street reconstruction

7:30 影像采集

12:00 影像采集

14:00 影像采集

19:30 影像采集

不同时段影响采集
Different time periods affect collection

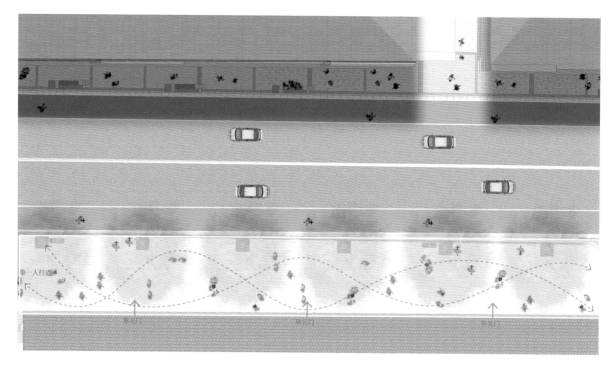

街道改造前
- 单元门直面街道
- 店招杂乱
- 部分一层开窗内部空间为公共厨卫
- 单车停放无序
- 人行道铺装单一
- 护栏功能性较弱
- 休息座椅缺少

街道改造前
Before street reconstruction

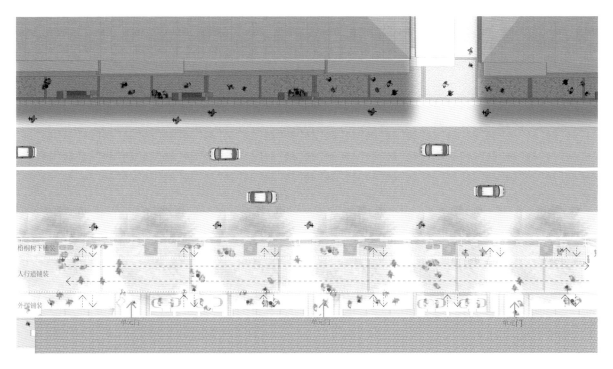

- 单元门增加铺装与花钵　街道改造后
- 店招统一
- 透光格栅美化一层界面
- 规划单车停靠点
- 划分外摆铺装、梧桐树下铺装、人行道铺装
- 护栏结合座椅与种植箱

街道改造后
After street reconstruction

效果图
Perspectives

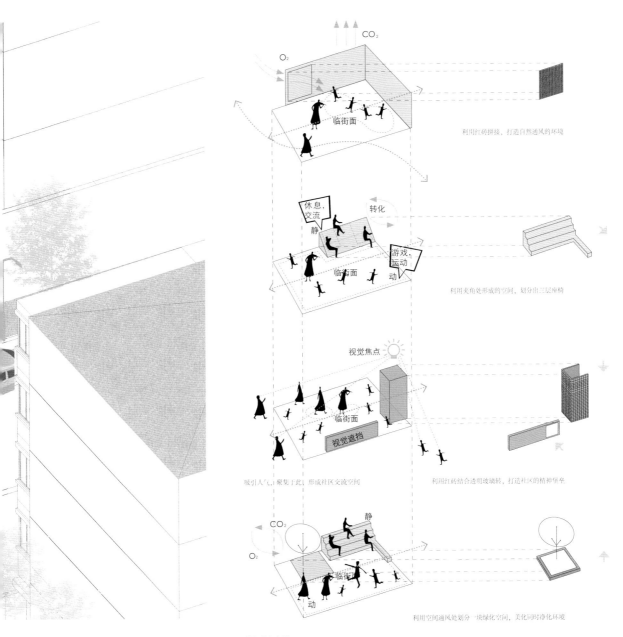

田林汇空间解构
Spatial structures

田林印记效果图
Tian Lin imprint effect diagram

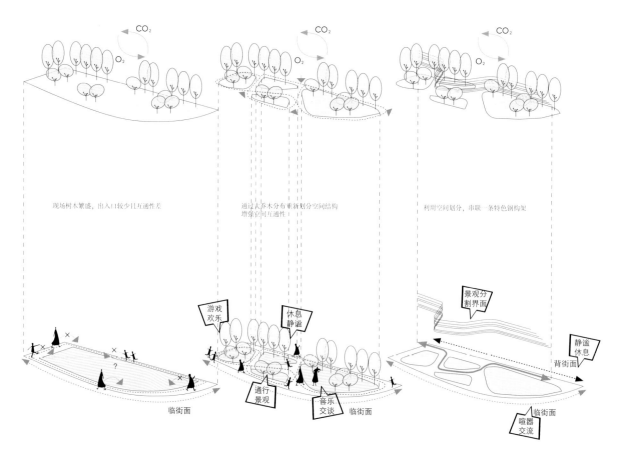

田林印记空间解构
Tian Lin mark space deconstruction

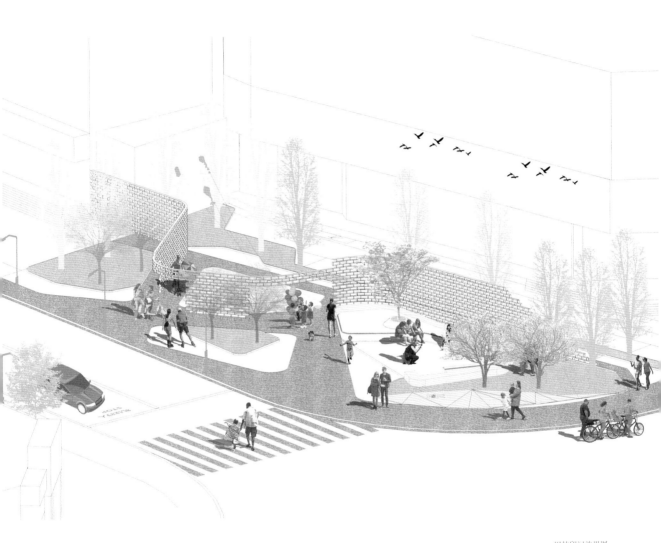

田林印记效果图
Tian Lin imprint effect perspective

187

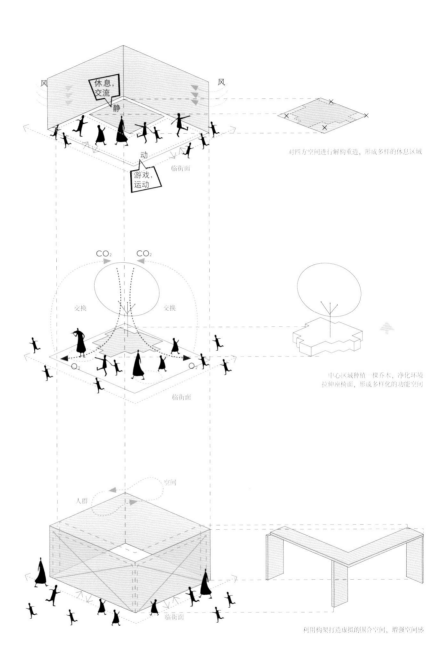

田林印象空间解构
Tian Lin impression space deconstruction

田林印象效果图
Perspective of Tian Lin impression

公园里的游憩时光 | 西安理想城
Tour Time In The Park
Ideal City XiAn

地点 | 陕西省西安市
规模 | 7.4万平方米
业主 | 西安万科企业有限公司

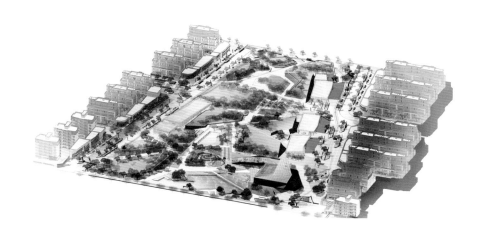

冰川肌理
Texture of glacier

优化组合
Optimum combination

列分控制
Optimum combination

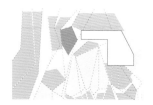

最终平面
Final plan

公园化社区商业
| 从纯粹的观察者走向真实的体验者

提升社区商业公园化的目的是为了通过景观设计来提升社区商业的综合价值和居民的生活质量。景观环境不只是人们欣赏的对象，更重要的是能优质地满足居住者的各项需求。理想城项目尝试提供一种氛围，让身处自然中的人们去想象自我与世界、自然的关系，将人们从纯粹的观察者变成真实的体验者。从形式到功能，体验真实的张力。

The purpose of promoting commercial parks in the community is to enhance the comprehensive value of community business and the quality of life of residents through landscape design.The landscape environment is not only the appreciation object of people, but should meet the needs of the inhabitants well.The Ideal City Project attempts to provide an atmosphere in which people in nature can imagine the relationship between the self, the world and nature, transforming people from pure observers to real experiencers.Experience the true tension from form to function.

Under the clouds meditate on the lawn

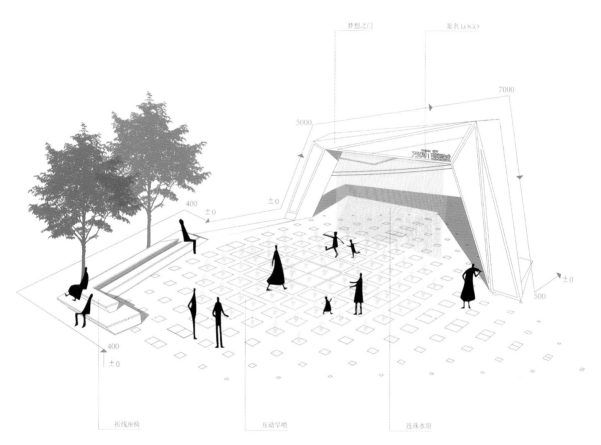

理想之门空间分析图
The analysis chart of the ideal gate space

理想之门广场
Square of Ideal-gate

夜晚音乐喷泉下的快乐时光
Happy hour at night under the music fountain

当当车站入口效果图
Perspective of Dangdang station entrance

市民公园效果图
Perspective of public parks

perspectives of pedestrian mall

穿梭于商业空间的当当车站
Dangdang station shuttling through the commercial space

当当车站构架
Frame of Dangdang station

城市界面的硬质驳岸与公园界面的软质驳岸
Hard revetment at city interface and soft revetment at park interface

下页 公园示范区剖面图
Section diagram of the park demonstration area on the next page

沉桥断面图
Section of sunken bridge

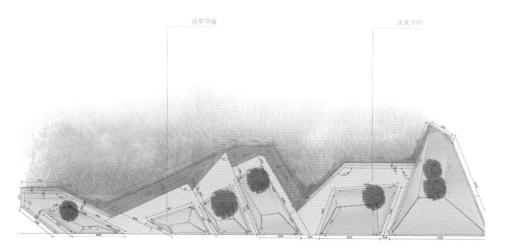

滨水平台平面图
Plan of waterfront platform

沉桥——水上五厘米的设计
Design of sunken bridge - 5cm above water

屋顶花园解构
Deconstruction of the roof garden

公园后场鸟瞰
Aerial view of the park's backfield

人的行为模式与植物的交融
The blending of human behavior patterns with plants

将自然的体验一直延续至售楼处的屋顶之上
The natural experience continues to the roof of the sales office

205

悬浮于水面的洽谈区
A negotiation zone suspended above the water

蜘蛛乐园
The spider paradise

昆虫主题的儿童活动场地有着多样的游戏体验
The insect-themed children's activity space has a variety of game experiences

成体系的游乐场地让孩子们乐在其中
The systematic playground makes children enjoy themselves

城市里的公园盒子 | 南昌博览城绿带公园
Park Box In The City
Guobo Green Park In NanChang

地点 | 江西省南昌市
规模 | 1.6万平方米
业主 | 绿地控股集团江西事业部

"蝶恋花"主题游戏体验关卡设计
| 关于空间设计的游戏

一块带状绿地希望能打破现状的模式,可以提供社区居民一个户外交流活动的绿地。空间游戏的构思也由此而生,造梦者以"蝶恋花"为主题创造了一个"花在明月蝴蝶梦"的主题游戏,并以三个关卡链接。带状绿地也将创造一个城市组团;住区无界的空间序列及模式,通过人在不同空间环境中的转换感受到一些有趣的、幸福的、放松的空间活动及游戏。创造一个城市、人、环境共享的舒适的活动交流空间。

A strip of green space hopes to break the status quo and provide a green space for community residents to communicate outdoors. The idea of a space game was born, and the dreamers created a "flowers amid the moon and butterfly dream" themed with "love of butterfly", with links of three levels. The strip of green space will also create an urban agglomeration; with unbounded spatial sequences and patterns in the residential area, through the conversion of people in different space environments, we can feel some interesting, happy, relaxed space activities and games. Create a comfortable space for communication between cities, people and the environment.

No.05

南昌博览城绿带公园

LANDSCAPE DESIGN ABOUT GUOBO GREEN PARK

2018
NAN CHANG

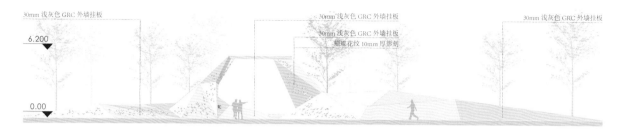

花蝴蝶的童话书
Story book of the butterfly

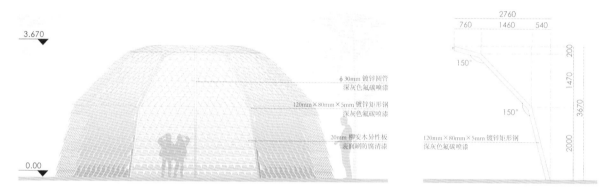

簇拥的茧亭
Cocoon pavilion

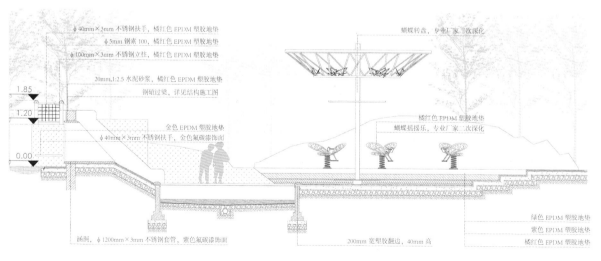

蝴蝶太阳花乐园
Paradise of butterfly sunflower

公园提供了可让各年龄阶段儿童安全玩耍的活动场地
The park provides a safe place for children of all ages to play.

主题游戏体验式关卡设计
Design of theme game experience level

书本折叠的童话书
Folding storybook

蝶形图案转模
Popular Science of Butterfly

蝴蝶科普故事
Butterfly popular science story

地景式的转折空间
| 既是一场亲近自然的精神之旅，亦是一场生动的科普故事

Breaking the cocoon into a butterfly is not only a spiritual journey close to nature, but also a vivid educational journey of popular science story. A broken line-shaped landscape-turning space is like a fairy tale book for children. It is a three-dimensional and vivid butterfly science knowledge after being unfolded. With the installation art, pattern, text, the butterfly from the egg to the butterfly is shown and the entrance square space is provided with new concrete material to present the outline of the book through folding. At the same time, the butterfly pattern is used to transfer the mold on the concrete to create a butterfly flying scene, focusing on the top.

破茧成蝶，既是一场亲近自然的精神之旅，亦是一场生动的科普教育之旅。一条折线形的地景式转折空间如同童话书，翻开来便是立体而生动的蝴蝶科普知识。设计师用装置艺术、图案、文字的方式，将蝴蝶从卵到蝶的过程展现出来，入口广场空间采用新型混凝土材料，折叠出书本的轮廓。同时在混凝土上用蝴蝶形图案转印模，营造出蝴蝶飞舞的景象，聚焦于顶端。

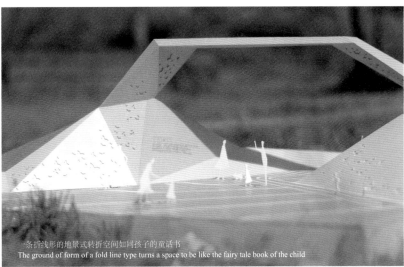

一条折线形的地景式转折空间如同孩子的童话书
The ground of form of a fold line type turns a space to be like the fairy tale book of the child

折线形的转折空间
A mansard landscape space

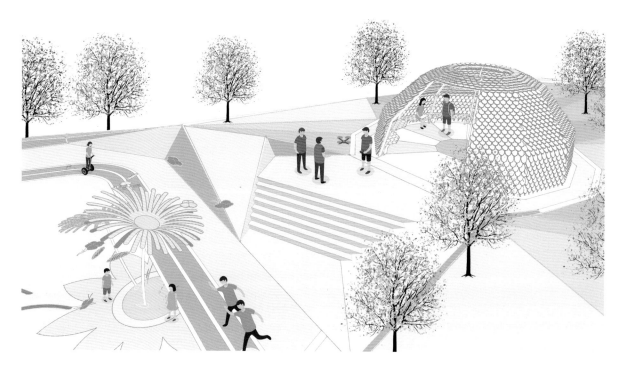

簇拥的茧亭
Cocoon pavilion

光与影的视觉焦点
| 通过这种有趣的互动,唤醒人们对自然探索的好奇心

The public landscape in the venue is shaped like a butterfly cocoon, located at the height of the site, with the best view of the landscape. This is the rest and exchange center of the whole area. People can sit on the bench in the pavilion and bamboo fence provides a more private space. The bamboo wood that can be flipped on the pavilion allows the people in the pavilion to enjoy the wonderful changes brought by the wind and light shadow. The unique patio design allows people to sit in the pavilion and look up at the sky. By means of damping device, the scale is flipped and the direction and intensity of light and wind are changed. This takes the cocoon pavilion as the medium and makes people participate in the interaction with nature. Through this interesting interaction, it can awaken people's curiosity about nature exploration.

场地中的公共景观以碟茧为造型,位于场地的制高点,拥有最好的景观视野,这是全区的休憩交流中心,人们可以在亭中长椅静坐,竹木的围挡提供了一个较为私密的空间。茧亭上可以活动翻转的竹木,使人能够享受到风和光影带来的美妙的变化。独特的天井设计让人安坐于亭内就能仰望天空。通过阻尼装置,带动鳞片翻转,改变着亭内光和风的方向与强度。这是在以茧亭为媒介使人与自然互动,通过这种有趣的互动,唤醒人们对自然探索的好奇心。

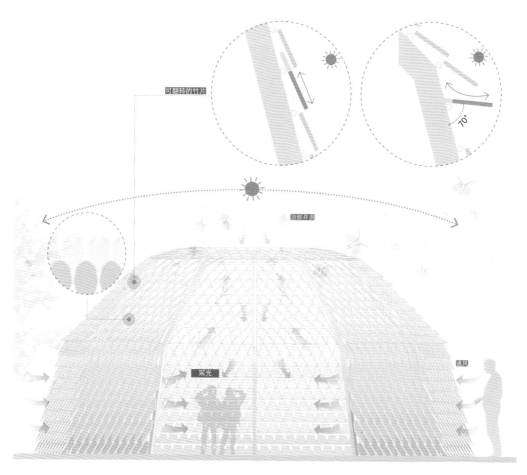

气候时钟
Clock of climate

捕捉太阳能量的有趣装置
给人们提供一个全天候，全季节的舒适空间

Relying on the curved shape, this introverted space creates a natural circle that captures the sun. The shape and orientation of the curved gallery makes it play the role of a "climate clock". The seat under direct sunlight can be found at any time during the day as long as people are willing to move. The curved cocoon pavilion has enough height to keep out the cold wind in the winter and creates a small shelter; on the contrary, in summer, people can find shady shadows according to the movement of the sun. This shape of the venue can provide people with an all-weather, all-season comfortable space.

依靠弧形的外形，给这个内向的空间创造了一个自然的捕捉阳光照射的圆圈，弧形廊架的外形和朝向使得它扮演一个"气候时钟"的角色，白天任何时刻只要人肯移动都能找到阳光直射的坐席。弧形的茧亭又有足够的高度来遮挡冬季冷风，创造了一个小小的庇护所；相反，夏日人们可以根据太阳的移动来找到阴凉的阴影。这样形状的场地可以给人们提供一个全天候，全季节的舒适空间。

捕捉阳光照射的向内空间
Capturing the inward sunlit space

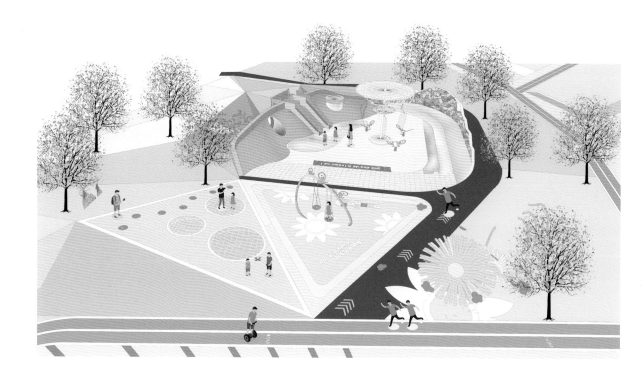

花朵装置
Installation of flowers

蝴蝶秋千
Butterfly swing

彩色跳床
Color trampolines

全龄场地
All age site

全年龄儿童游戏场地
| "久在樊笼里，复得返自然"的一次破茧成蝶的向往

The park provides an event venue for children in all ages to participate in safety games. Through the reinterpretation of the elements of the butterfly and butterfly garden, these are reflected in the game installation and the landscape art, with colorful colors, flower-shaped installations and the butterfly-shaped instruments. These make the children who play in this place for children of all ages seem to be in the garden surrounded by butterflies.

The fairy-tale book-like landscape art, the cocoon pavilion that interacts with the light and shadow breeze, and the garden-like children's space all guide people to explore nature actively. It is pressure release in the inner heart of adults, learning of children for the world and people's desire to return to nature... through this space design game, the longing for cocoon-break for "after staying in the cage for a long time, returning to nature" is reflected.

公园提供了可让各个年龄阶段儿童安全游戏的活动场地，通过对蝴蝶及蝴蝶生活的花园等元素的重新阐释，将这些体现在游戏装置以及地景艺术上，缤纷的色彩，花朵造型的装置与蝴蝶造型的器械，使在这片全龄儿童活动场地嬉戏的儿童仿佛置身蝴蝶环绕的花园之中。童话书般的地景艺术、与光影微风互动的茧亭、花园般的儿童空间，都在引导着人们主动探索自然，是成人对心灵的解压，是儿童对世界的学习，是人们对重归自然的渴望……通过这场空间设计的游戏，寄托着"久在樊笼里，复得返自然"的一次破茧成蝶的向往。

花园般的儿童乐园引导人们探索自然
A children's garden leads people to explore nature

儿童仿佛置身蝴蝶环绕的花园之中
Children are placed in a garden surrounded by butterflies

上升的螺旋引力 | 南昌博览城庆典公园
Rising Spiral Gravitation
Celebration Park In NanChang

地点 | 江西省南昌市
规模 | 9.7 万平方米
业主 | 绿地控股集团江西事业部

斐波那契螺旋线
| 大自然中的斐波那契数列螺旋

螺旋让信息能够紧密地结合其中，允许其他微粒在一定的间隔处与它相结合。例如，DNA 的双螺旋结构允许进行 DNA 转录和修复。衍生到场地空间各个功能的黏合、场景空间的相互作用。庆典公园基于国博会展中心的建筑设计语言向外扩展，衍迤出斐波那契黄金螺旋的设计语言形式。这种方式可最大化引导会展活动人员疏散以及提供充足的等候缓冲空间。

The helix allows information to be closely integrated, allowing other particles to be combined with it at certain intervals. For example, the double helix structure of DNA allows for DNA transcription and repair.- Cohesion of various functions between the space of the venue and the interaction of the scene space. The Celebration Park extends on the basis of the architectural design language of the National Museum Convention and Exhibition Center to develop the design language form of the Fibonacci gold spiral. This approach can guide evacuation of personnel involved in convention and exhibition activities to the greatest extent and provide adequate waiting buffer space.

NO.06 南昌博览城庆典公园

LANDSCAPE DESIGN ABOUT CELEBRATION PARK

2018 NAN CHANG

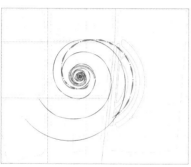

建筑边界与场地关系

城市人流与场地关系

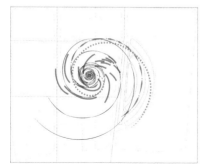

场地内部关系

场地外部关系

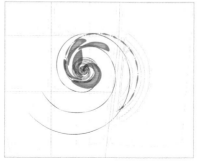

场地铺装纹理

场地高差关系

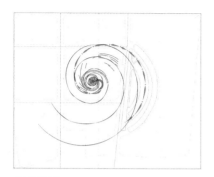

场地构筑艺术墙

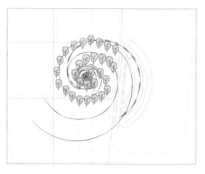

场地种植纹理

场地空间分析图
Site space analysis diagram

Analysis of the relationship between the ground and the Fibonacci gold thread on the right page

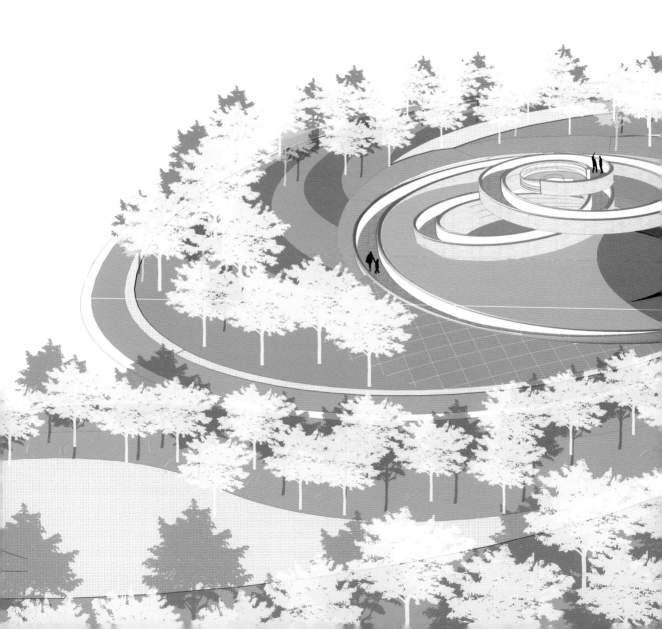

环形构架与场地高差完美结合
The ring structure is perfectly integrated with the site height difference

弧形的场地纹理形成空间的视觉引导
The curved site texture forms spatial visual guidance

环形构架模型外观
Appearance of the ring architecture model

硬质景观与植物结合
The hard landscape blends with the plants

环形构架模型细节
Detail of ring architecture model

"环形"设计完美结合场地地形高差，底层的桥梁步道既满足是人行同时也能营造高低点游览的不同视角

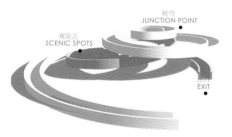

衍生到场地空间之间各个功能的黏合、场景空间的相互作用。庆典公园基于国博会展中心的建筑设计语言向外扩展，演绎出斐波那契金螺旋的设计语言形式。这种方式可最大化引导会展活动人员疏散以及提供充足的等候缓冲空间。

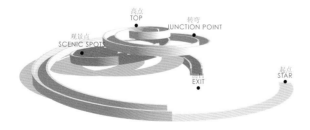

螺旋让信息能够紧密地结合其中，允许其他微粒在一定的间隔处与它相结合。

螺旋观景构架空间解构
Spatial deconstruction of spiral viewing frame

衍生到场地空间之间各功能的契合
Derived to the adhesion of various functions between the site space

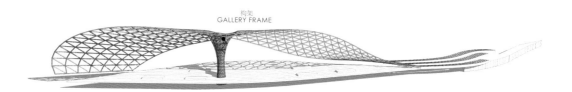

"编织网"状设计型的廊架体现曲线延伸之感,同时在功能上能为周边休憩游人提供休息遮阳的作用。

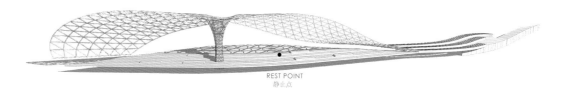

廊架结合木质阶梯状休憩座椅的设计既满足游人休憩需求,同时完美结合廊架延伸的造型。

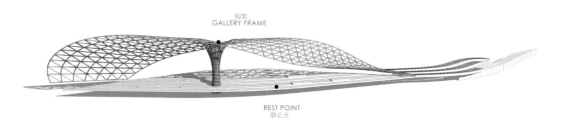

整个场所由廊架和木质阶梯状休憩空间构成,为整个场所提供灵动的感觉。

运动场休憩廊架效果图
Perspectives of the playground recreation gallery frame

廊架模型细节
Details of the gallery frame model

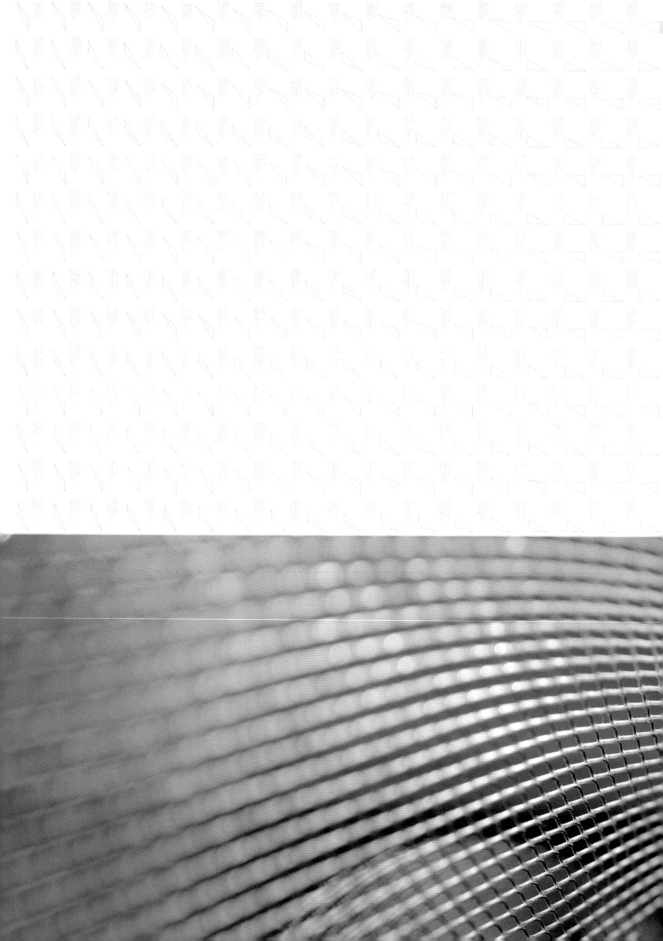

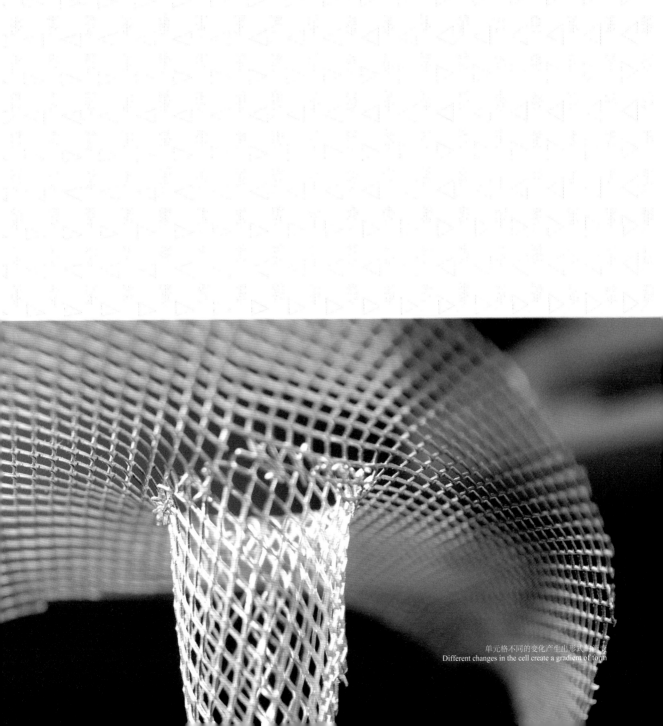

单元格不同的变化产生出形式的渐变
Different changes in the cell create a gradient of form

浮动的森林 | 南昌安南小镇玻璃花房
Floating Forest
Annan Town Glass Greenhouse Nanchang

地点 | 江西省南昌市
规模 | 900平方米
业主 | 绿地控股集团江西事业部

探索人与自然的关系
| 神秘雨林之旅,感受自然之肺

几百年来,人类对大自然一直存在以人为中心的傲慢态度,砍伐森林带来的大自然对人类的惩罚一直在发生,全球每分钟就有25公顷热带雨林消失。自然是人类的生命层,保护自然、保护森林是紧迫要做的事情,也是我们温室设计的初衷。以"浮动森林"为主题,还原热带雨林风貌,重视热带雨林的"呼吸"功能,树木通过光合作用,吸收二氧化碳,向大气中补充氧气,通过人体验不同高度、视角,营造神秘的雨林环境,入内由热带雨林而起,藤蔓密织,转而到厅,厅内奇树古木,珍花稀草,悬葛垂萝,其右浮桥,蜿蜒而上,历经"林湾洞瀑溪涧岸堑"八景而回,体验有"地球之肺"之称的热带雨林环境,感受大自然的消失与存在,孤立与联系。

For centuries, human beings have always had a human-centered arrogance towards nature. The punishment of nature brought about by deforestation has been happening, and 25 hectares of tropical rain forests disappear every minute in the world. Nature is the life layer of mankind. It is urgent to protect nature and forest and it is also the original intention of our greenhouse design. Take "floating forest" as the theme, it is required to restore the tropical rainforest scenery, attach importance to the "breath" function of the tropical rainforest. Through photosynthesis, trees absorb carbon dioxide, replenish oxygen to the atmosphere. Through the experience of different heights, perspectives, it aims to create a mysterious rainforest environment. From the rain forest inside, the vine is densely weaved. When we walk to the hall, there are strange trees and ancient woods, rare flowers, rare grasses, drooping vines. The floating bridge on the right side wriggles up to go through eight sceneries including "forest, guff, cave, waterfall, stream, brook, bank and chasm". Therefore, we can experience the tropical rain forest environment that is honored as "lung of the earth" and feel the disappearance and existence, isolation and connection of nature.

No.07

南昌安南小镇玻璃花房

LANDSCAPE DESIGN ABOUT ANNAN TOWN GLASS GREENHOUSE

2018
NAN CHANG

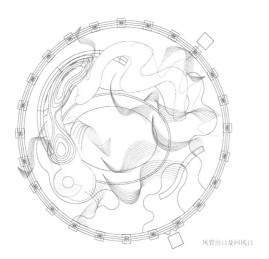

风管出口及回风口

风管出口
回风口倒置系统

降温：埋藏式风管送风，恒常的土地热惰性保持着空气的凉爽。低温空气穿过内部空间，通过生成空气对流将热空气抽出

保温：通过日照辐射及夹层高温空气，实现了温室效应。在冬季，通过关闭通风系统来延迟热辐射。由于土壤热惰性，在基础中积蓄了热量，可被回收利用

风管系统和水体面积：潜热降温
Air dust system and water area: Latent heat and cooling

风管系统：回收蓄热
Air duct system: heat recovery and storage

消除：防晒。在全年中日照最盛的月份，内部的移动式遮阳机械系统减少了直接辐射

能量：表皮蓄热缓和了室外温差，控制室内温度

减弱：在夏季的月份中避免直接辐射，这得益于移动式遮阳系统。通过边缘引风风管进行通风

积蓄：在双层膜表皮中；在室内空腔中，在热惰性的夹层中

风管系统和水体面积：潜热降温
Air duct system and water body area: latent heat cooling

风管系统：回收蓄热
Air duct system: heat recovery and storage

244

夏季日照分析
Summer sunshine analysis 通过热带植物以及绿植墙的布置对开敞空间进行一定程度的遮挡 SUMMER SUNSHINE ANALYSIS

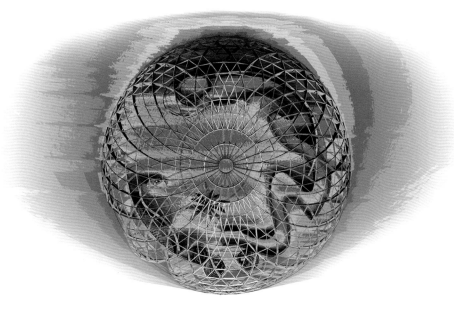

冬季日照分析
Winter sunshine analysis 通过设计尽可能地使开敞空间获得更多的日照时间 WINTER SUN EXPOSURE ANALYSIS

建筑龙骨

绿植墙

浮动飘带

玻璃花房立面图
Elevation of glass greenhouse

瀑布

玻璃扶手

生态叠水

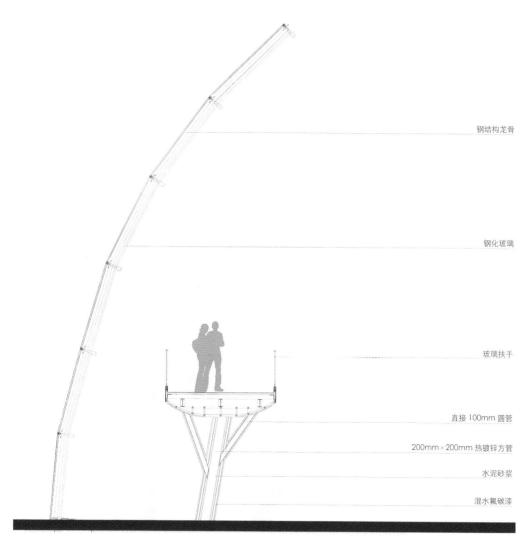

半空廊桥与玻璃穹顶剖面图
Section of the air-bridge and the glass dome

钢结构龙骨

钢化玻璃

玻璃扶手

直接100mm圆管
200mm×200mm热镀锌方管
水泥砂浆
混水氟碳漆

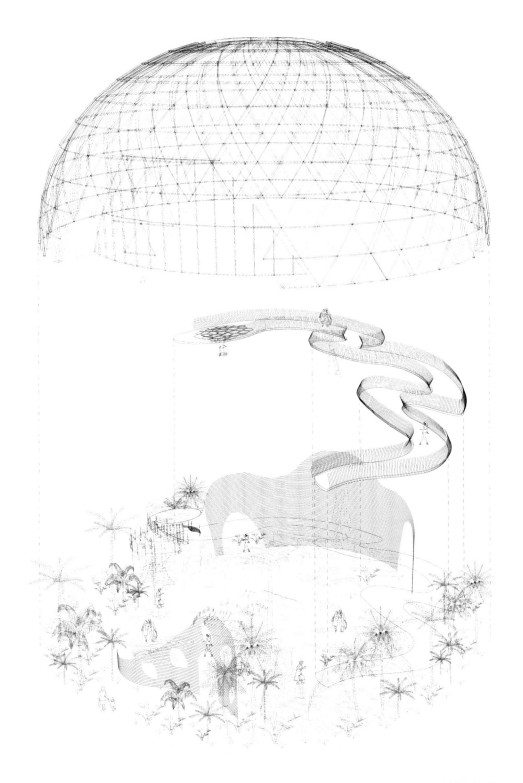

玻璃花房空间解构
Spatial structures of glass greenhouse

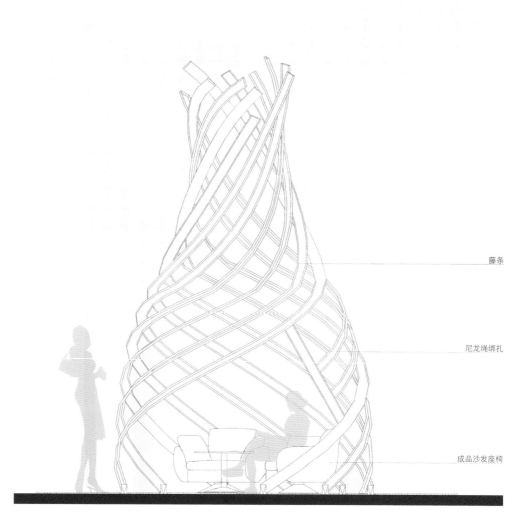

森巢立面图
Forest nest elevation

藤条

尼龙绳绑扎

成品沙发座椅

森巢——绿色丛林中的鸟巢
Forest nest – bird nest in a green forest

半空廊道与玻璃穹顶
Skywalk and glass dome

大地美术馆 | 南昌红土遗址公园
Earth Art Museum
Red Earth Heritage Park Nanchang

地点｜江西省南昌市
规模｜19.8万平方米
业主｜绿地控股集团江西事业部

不朽荒漠，大地美术馆
| 过去现在与未来，一场流动的时空变幻之旅

在红土的地表下，是这片红土的地质构造和背后的地质学意义，公园是第四纪网纹红土组成的红色荒漠景观为主要特色的地景公园，场地生有大片的马尾松林和其他植物群落。集大地景观，自然体验，生态修复，科普认知，休闲放松为一体，内主环路和不同宽度的次级游园路连接三个出入口，十二个景点依次展开，每个景点不同特色，方便人们去探寻和了解自然变换之美。红土地貌较为脆弱，雨水不易下渗，地表径流引起坡面冲刷，致水土流失严重，形成植被稀疏，沟壑纵横的红色荒漠景观。红土地貌随着时间而变化，对研究古地理环境，新构造运动及古冰川作用有极为重要的研究意义。设计提倡教育性与参与性，使文化和自然环境融合，关注生命与土地的和谐相处，体现生态、美学、社会三维价值体系。

Under the surface of the red earth is the geological structure of the red earth and the geological significance behind it. The park features a red desert landscape composed of quaternary reticulated red earth, with a large area of masson pine forest and other plant communities. It integrates landscape, natural experience, ecological restoration, popular science knowledge, leisure and relaxation. The main ring road and the secondary garden road of different widths are connected with three entrances and twelve scenic spots spread out in turn. Each scenic spot has different characteristics, which is convenient for people to explore and understand the beauty of natural transformation. The red soil landform is relatively fragile, the rain water is not easy to infiltrate, and the surface runoff causes slope erosion, resulting in serious soil erosion, forming a red desert landscape with sparse vegetation and ravines. The laterite landform changes with time, which is of great significance to the study of paleogeographic environment, neotectonic movement and paleoglaciation. The design advocates education and participation, integrates culture and natural environment, and pays attention to the harmonious coexistence of life and land.

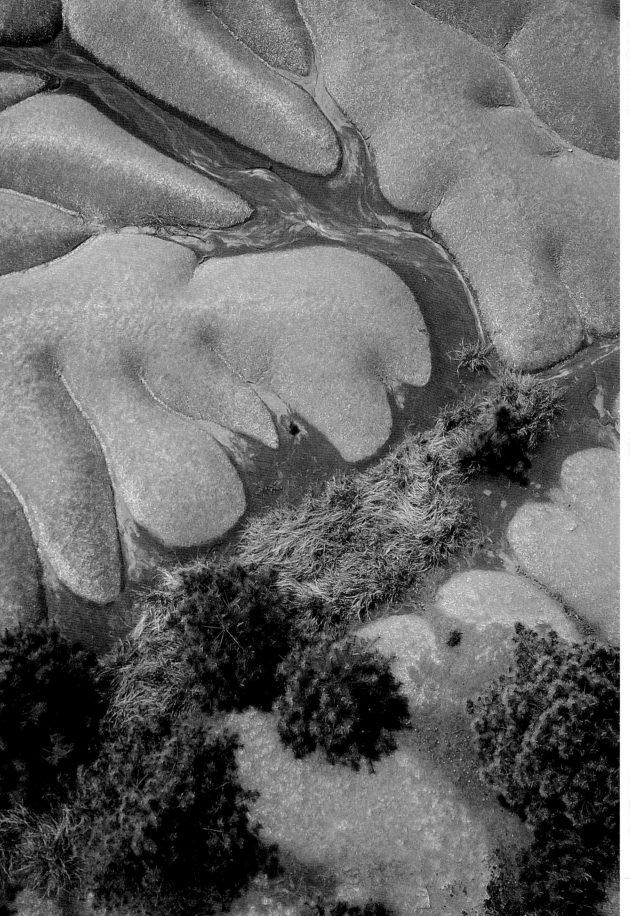

NO.03 南昌红土遗址公园 LANDSCAPE DESIGN ABOUT RED EARTH HERITAGE PARK 2018 NAN CHANG

2013.01　　　　　　　　　　　　　　　　　　　　　　2016.12

卫星图变化
Change of satellite imagery

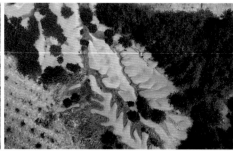

鸟瞰图　　　　　　　　　　　红土地被层次　　　　　　　　　红土细节

场地现状调研
Investigate of site situation

2018.02

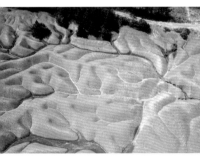

红土高度

雨水冲刷痕迹

周边地貌

场地现状调研
Investigate of site situation

径流分析
Runoff analysis

现状
Current situation

五年后
Five years later

十年后
Ten years later

2 年　一至二层群落（先降物种入驻）
2 years 1-2 story communities (descending species first)

（乔木 + 地被）
（马尾松、刺槐等）

10 年　二至三层群落
Two - to three-layer community for 10 years

（乔木 + 亚乔木 + 地被）
（壳斗科、樟科、山茶科、木兰科、刺槐、草本层莎草科、禾本科）

30 年　三层群落
Three-layer community for 30 years

（乔木 + 亚乔木 + 地被）
（壳斗科、樟科、山茶科、木兰科、刺槐、地被常绿杜鹃属、草本层、莎草科、禾本科等）

50 年　四层群落
Three-layer community for 30 years

（乔木 + 亚乔木 + 地被）
（青冈、石栎、壳斗科、樟科、山茶科、木兰科、刺槐、常绿杜鹃属等）

红土公园入口模型日照分析
Sunlight analysis of entrance of Red Earth Park model

粗筛土
Coarse sieve soil → 细筛土
Fine screen soil → 加石灰用水搅拌混合
Add lime and mix with water

→ 区别红土黄土分层，加混凝土纤维增强
Distinguish red loess layered, reinforced with concrete fiber   人工运输
Artificial transportation

→ 气动锤粗夯，人工细夯
Pneumatic hammer rough tamper, manual fine tamper → 夹板固结保护
Splint consolidation protection

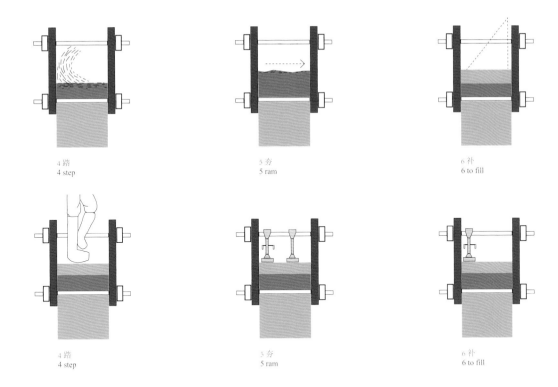

4 踏
4 step

5 夯
5 ram

6 补
6 to fill

4 踏
4 step

5 夯
5 ram

6 补
6 to fill

公园展示区入口夯土墙
Rammed earth wall at entrance of park exhibition area

263

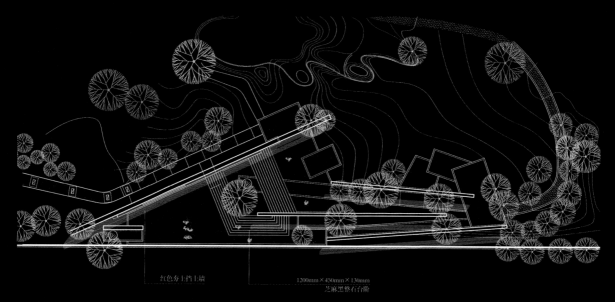

公园示范区入口平面图
Plan of entrance of park demonstration area

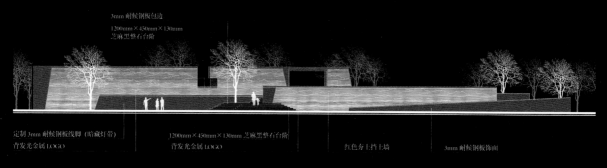

公园示范区入口立面图
Elevation of entrance of park demonstration area

公园入口夯土景墙外观
Appearance of rammed earth landscape wall at the entrance of the park

EARTHEN WALL

公园入口日照及交通动线分析
Analysis of sunshine and traffic line at the entrance of the park

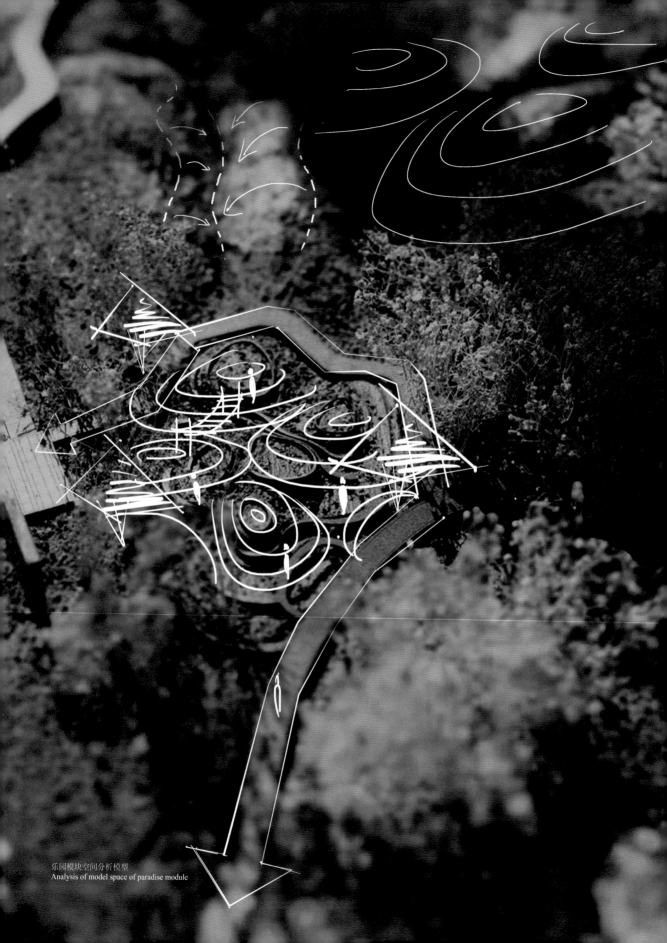

乐园模块空间分析模型
Analysis of model space of paradise module

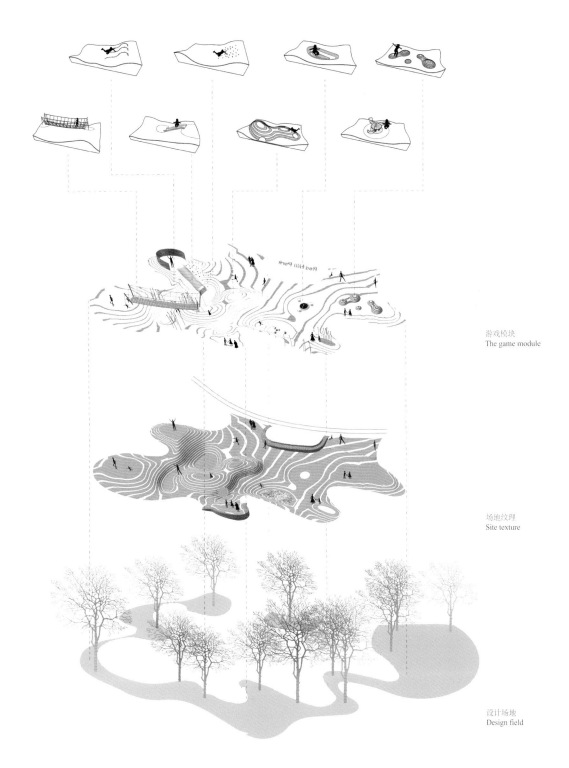

游戏模块
The game module

场地纹理
Site texture

设计场地
Design field

阳光下的松果草甸
Inecone meadows in the sun

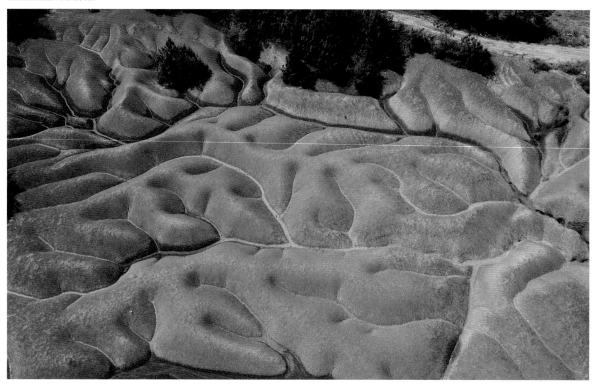

自然的艺术肌理
Natural art texture

轻架空的木栈道 —— 低影响设计与建造
Trestle walkway - low impact design and construction

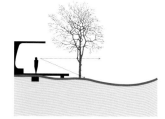
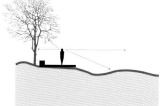
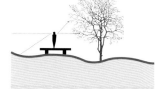
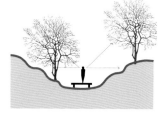

栈道观望台视线分析
View line analysis of the plank road

栈道观望台场地模型
Site Model of Platform on Trestle Road

寻找和大海对话的空间语境 | 三亚 JW 万豪酒店
Find The Spatial Context Of Dialogue With The sea
JW Marriott Hotel Sanya

地点 | 海南省三亚市
规模 | 1.6 万平方米
业主 | 鲁能集团三亚公司

用风捕捉大海的形状
| 随风而行，随遇而安

身在海边，如果不能感受到海风的吹拂，如果视线不能延续直到海面，这样的场所就不具有临海的特质，我们就需要重新找到和大海对话的方式。这是一个酒店因品牌升级而改造外环境的景观设计，原场地虽临海但现状空间形态没有形成和海景的联系，植物甚至阻隔了看海的视线，为了让人和海形成互动，设计以分析海边的风、光环境为基础，创造了顺应海边自然要素的场所形态，为了把视线引向海面，设计在场地和海面之间设置了镜面水景，并在水面之上创造了一处艺术构架，构架的设计来源于海风的形态，仿佛凝固了的海风，本来静止的构架也具有了动态的感觉，在这里抽象的海被转化成变幻的光、影、风、形，在人的坐、卧、行间仿佛重新看到了大海的样子。

Being at the beach, if we cannot feel the blowing of the sea breeze and the line of sight cannot continue till the sea, such a place does not have the coastal characteristics and we need to find a way to talk to the sea again. This is a landscape design that transforms the external environment of the hotel due to brand upgrade. Although the original site is not near the sea, the current spatial form does not form a connection with the seascape. The plants even block the sight of the sea. In order to make people interact with the sea, it is designed to create a form of place that conforms to the natural elements of the seaside on the basis of analyzing wind and light environment of the seaside. In order to divert the line of sight to the sea surface, a mirrored waterscape is set between the site and the sea surface, and an artistic structure is created above the water surface. The design is derived from the shape of the sea breeze, as if the sea breeze was solidified. The original static structure also has a dynamic feeling. Here, the abstract sea is transformed into a changing light, shadow, wind, shape. It seems as if I saw the sea again during sitting, lying, and walking of people.

NO.09

三亚 JW 万豪酒店

LANDSCAPE DESIGN ABOUT JW MARRIOTT HOTEL

2018 SANYA

灯光模拟海洋生物发光的自然现象，让海的意境更加强烈
The lighting simulates the natural phenomenon of Marine bioluminescence, making the sea more vivid

透过艺术构架,远望海面远山
Through the artistic framework, we can see the sea and the mountains

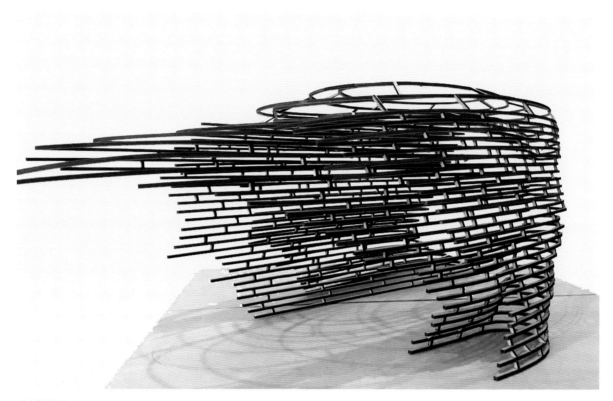

艺术构架模型
Art frame model

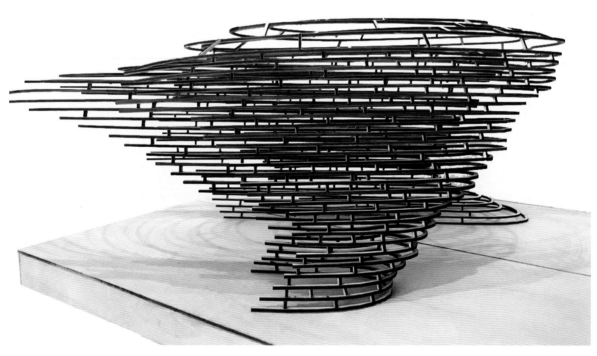

模拟海风吹拂，随风舒展的形态
The simulation sea breeze blows, the wind stretches the shape

顶视图
Top view

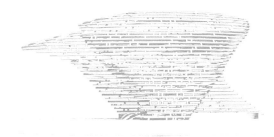

侧视图
The strakes

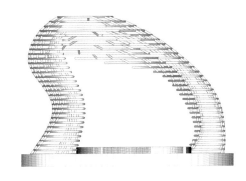

正视图
Front view

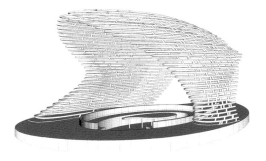

透视图
Perspective

平面的形态让海岸和建筑融为一体
The form of the plan allows the coast to merge with the building

抽象的"海"被转化成变幻的光、影、风、形
The abstract "sea" is transformed into light, shadow, wind and form

在空间中感受大海的模样
Feel the sea in space

静止的水面倒映着廊架
The still water reflects the gallery

灯光烘托了夜晚的氛围
The lighting sets the mood for the night

顺应自然要素的场所形态
Conform to the site form of natural elements

透过水面的倒映使静态的空间有了动态的意味
The static frame is reflected by the water surface, giving the static space a dynamic meaning

静水面和浮动的"小岛"
Still waters and floating "islets"

顺应自然要素的场所形态
Conform to the site form of natural elements

前卫与本真的对话 | 郑州长安古寨
Avant-garde And Genuine Dialogue
Zhengzhou Chang'an Ancient Village

地点 | 河南省郑州市
规模 | 30 万平方米
业主 | 河南建业集团

黄土记忆
乡土景观的城市更新态度

新与旧不期而遇，我们应当抱以怎样的态度？
当本真质朴的材质和新的形式碰撞时，激起我们对生活真正内涵的思考也唤醒我们对城市快速发展的担忧。
如果，新的空间，延续有过去的记忆。老的场所，包容有新的生活。人和其他物种共生，人与人之间是友善和解，这或许便是未来的理想城市。长安古寨原是中原地区特有的黄土沟壑地貌聚落形态，在城市发展中遇到势在必行的开发，用什么样的方式能够充分尊重原生态同时又能融入现代人的生活？我们怕这种城市再生变成一种仿古的假象，又希望让古寨在发展中受到保护和延续，景观设计从研究当地黄土聚落的景观构成形态出发恢复古寨原有乡土美学的完整性，完全使用乡土的构建素材和工艺工法，从而保持乡土美学的真实性，让古寨回归朴素纯真。新的生活方式也将和当地非物质文化遗产共存。因新的城市功能诉求古寨也包含几处新建部分，这些部分将运用新的建造手法和老建筑真诚对话。古寨的再生不再是仿古的假象，而是真诚而质朴的新老时空对话，乡土美学的真实性将被保持，先进的意识形态也会被鼓励，这是一场返璞归真的生活回归之旅。

What attitude should we take when the new meets the old?
When the primitive material and the new form collide, it will arouse our thinking about the true meaning of life and awaken our concerns about the rapid development of the city. If The new space continues the memory of the past. Old places are inclusive of new life. People and other species can co-exist. It is friendly reconciliation between people. This may be the ideal city of the future. Chang'an Ancient Village was originally a unique form of loess gully in the Central Plains. It encountered imperative development in urban development. In what way can the original ecology be fully respected and can it be integrated into the modern life? We are afraid that this kind of urban regeneration will become an antique illusion, and we hope that the ancient village will be protected and extended in the development. The landscape design will restore the integrity of the original rural aesthetics of the ancient village from the study of the landscape composition of the local loess settlement. The construction materials and craftsmanship of the native land are completely used, so as to maintain the authenticity of the local aesthetics and let the ancient villages return to simplicity and innocence. The new way of life will also coexist with local intangible cultural heritage. The ancient village also includes several new parts because of the new city functions and they will use the new construction techniques to make a dialogue with the old building. The regeneration of ancient villages is no longer an illusion of archaism, but a sincere and simple dialogue between the new and old times and spaces. The authenticity of local aesthetics will be maintained and the advanced ideology will be encouraged. It is a journey of life regression for returning to nature.

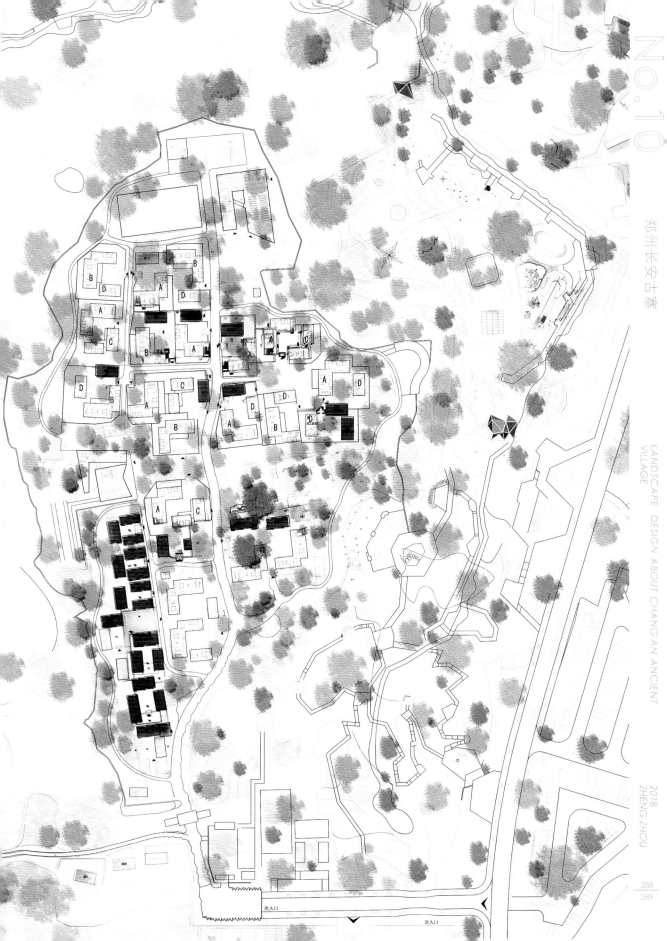

游客服务中心平面图
Visitor service center plan

游客服务中心透视图
Visitor service center perspective

游客中心的"新建筑"与古寨的"古建筑"隔空对话
The "new building" of the visitor center talks to the "old building" of the ancient village

游客中心夜景
Night view of the visitor center

游客中心鸟瞰
A bird's eye view of the visitor center

古寨赋予新的功能，新建筑将运用乡土材料和工艺手法
The ancient village is endowed with new functions, and the new building will use local materials and techniques

景观延续当地黄土聚落的构成手法，恢复古寨原有乡土美学的完整度，完全使用乡土的素材工艺
The landscape continues the composition of local loess settlements, restoring the integrity of the original local aesthetics of the ancient village, and completely using local materials and techniques

访客中心虽为新建筑，但其底层延续周边民居材质。屋顶的折线形态与民居的天际线连为一体，是对传统民居的充分尊重和真诚对话
Although the visitor center is a new building, its ground floor continues the material of surrounding residential buildings. The broken line form of the roof is connected with the skyline of the residence, which is a full respect and sincere dialogue to the traditional residence

从访客中心二层室内看向古寨建筑群落
View the ancient village building community from the second floor of the visitor center

如大树一样存在的景观 | 临沂生命织树
Existing Landscape Like The Tree
Tree of Life In Linyi

地点 | 山东省临沂市
规模 | 20.3万平方米
业主 | 旭辉银盛泰集团

生命织树,编制童梦空间

当人工建造的构架能如树冠一样遮阴和透光,构架上结出果实,孩子在上边玩耍,生活就在这大树下慢慢生长

小时候庭院的大树下边,总是吊着一个摇荡不停的秋千,那一根细细的绳子给了我们无尽的欢乐,坐在秋千上看大树结出果实,看树叶随风飘落,夏天我们沐浴在大树的光影之下,晚上看着天上的星星听奶奶讲故事,一棵树就是一个家的空间。

如果,我们创造一种空间,让它像大树一样存在。构架能够如树冠一样遮阴和透光,等大树结出果实孩子可以在上边玩耍,生活就可以在大树下生长。因为如"大树"一般的构架我们能和家人欢聚,在这能结交更多的朋友,那么没有真树也有自然。

When we were young, there was always a swing hanging under a big tree in the courtyard, and a thin rope gave us endless joy. Sitting on the swing, we watched the trees bearing fruit, the leaves falling with the wind. In the summer we were bathed in the light and shadow of the trees. Looking at the stars in the sky at night, we listened to Grandma's story. A tree is a home space.

If we create a space, let it exist like a tree. When the frame can be shaded and penetrated like a crown, after fruiting on the frame, children can play in it and life grows under the tree. We can unite with our family because of the "tree" structure. We can make more friends here. Then there is no real tree and nature.

NO.11

临沂生命织树

LANDSCAPE DESIGN ABOUT TREE OF LIFE

2018
SHAN DONG

生命织树透视图
Perspective of life woven tree

生命织树草图构思
Life weaves sketches

| 太空舱 | 网袋 | 秋千 |
| Space capsule | Mesh bag | Swing |

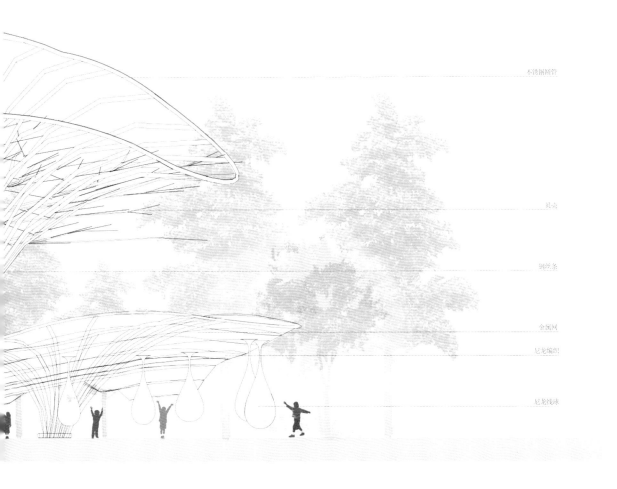

不锈钢圆管

贝壳

钢丝条

金属网
尼龙编织

尼龙线球

构架模拟大树自然生长的状态,悬吊的贝壳风铃随风飘摇,在树下倾听风的声音
The frame simulates the natural growth state of the tree. The hanging shell wind chimes sway with the wind and listen to the sound of the wind under the tree

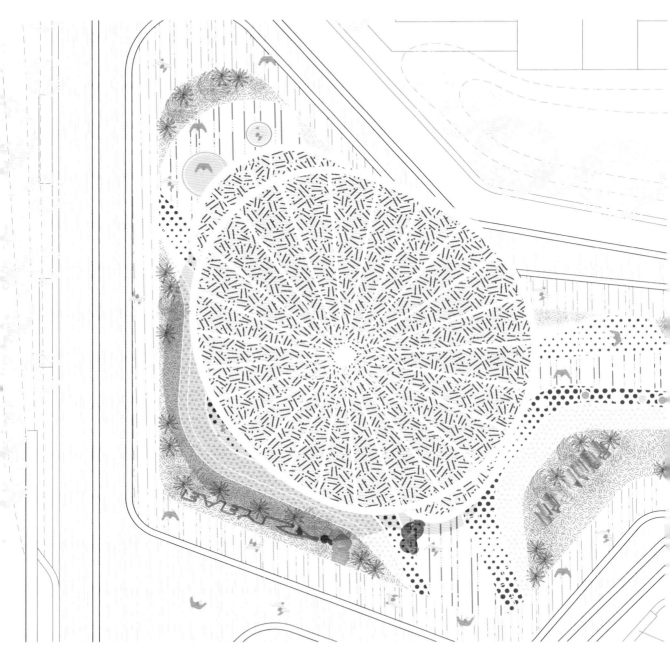

生命织树平面图
Plan of the tree of life

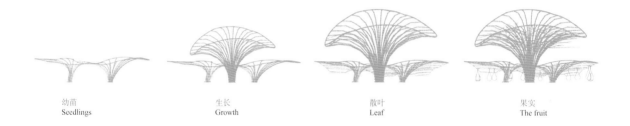

| 幼苗 | 生长 | 散叶 | 果实 |
| Seedlings | Growth | Leaf | The fruit |

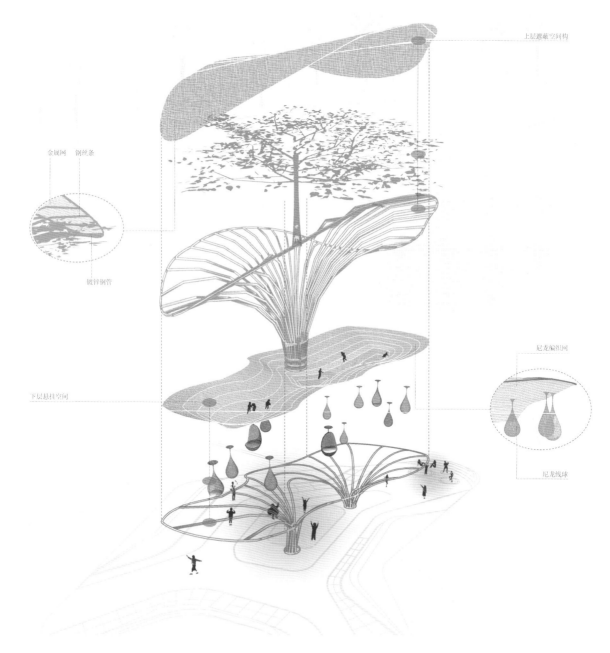

生命织树空间解构
The spatial deconstruction of the woven tree of life

大树构架既是遮阴挡雨的棚架又是儿童游戏的器械,同时也是一处交友聚会的复合式多功能场所
The tree framework is not only a shelter from the rain and the scaffolding is also a game equipment for children, but also a compound multi-functional place for friends.

"编织"是自然存在的最普遍的形式，在这里用编织的方式提供给人们最轻柔的触感
"Weaving" is the most common form of the most natural existence. Here, weaving is used to provide people with the most gentle touch.

网袋编织方法示意
Net bag knitting method

一个小小的网袋给了孩子们无尽的欢乐
A small net bag gave the children endless joy

孩子们在"大树"上玩耍
The children were playing on the big tree

地域文化的现代回归 | 昆明樾府
Modern Return Of Regional Culture
Kunming Orlental Mansion

地点 | 云南省昆明市
规模 | 5405平方米
业主 | 中南集团

在田埂上漫步
Walk along the ridge of the field

大地的调色盘
The earth's palette

瓦的海洋
Of the ocean

共生田园的回归之旅
| 大地艺术的在地诠释

昆明地域的特色,是人?是建筑?是景观?或许都不是。
设计中我们一直想找到一个具有在地性特征的意向,建筑、景观、室内分别寻找各自的地域特色而不得其要,其实一直忽略了它们之间的内在联系其实才是真正具有在地特色的内涵,云南的人、田、屋、山是整体存在的,它们之间存在无间的联系,那么云南的地域意向就是"共生的大地"意向,设计将"共生大地"的意向通过大地艺术的手法转换成景观建筑的整体呈现,当人被置身于一片屋瓦的海洋里看到大地和建筑连为一体时,似乎感受到这一似曾相识的共生的意向。在这个项目中我们思考了建筑和景观的关系,思考了景观的在地性诠释,思考了人和环境的关系,设计工作很多时候被专业板块的划分所限制,但实际上人、建筑和环境从来都不是孤立存在的。

The characteristic of Kunming area, is person? Architect? Landscape? Maybe neither. In the design, we have always wanted to find an intention with local characteristics. We search building, landscape and indoor to find their respective regional characteristics, but end up with nothing. In fact, it has been long ignored that the internal relationship between them is truly the connotation with local characteristics. People, fields, houses and mountains in Yunnan exist as a whole and there is ceaseless connection between them. Then regional intention of Yunnan is "symbiotic earth" intention. In the design, "symbiotic earth" intention is converted into the overall presentation of landscape architecture by the means of earth art. The idea of this familiar symbiosis seems to be felt when one is placed in a tiled ocean and the earth and architecture are joined together. In this project, we think about the relationship between architecture and landscape, the local interpretation of the landscape, and the relationship between people and the environment. Design is often limited by the division of professional sections, but in fact people, architecture and environment never exist separately.

No.12 昆明樾府

LANDSCAPE DESIGN ABOUT ORIENTAL MANSION

2018 KUN MING

从民居中提炼的木构架支撑的形态
The form of wooden framework support extracted from folk houses

梯田与山门是场地记忆中的属性
Terraces and gates are attributes of the site memory

入月门立面图
Moon gate elevation

入月门外观
Entry gate appearance

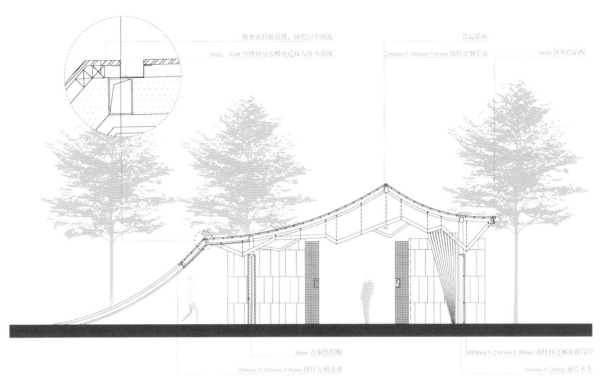

入月门立面
Moon gate elevation

入月门立面
Moon gate elevation

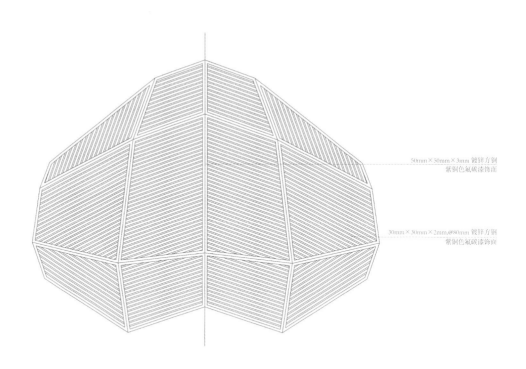

商山阁平面图
The plan of the pavilion

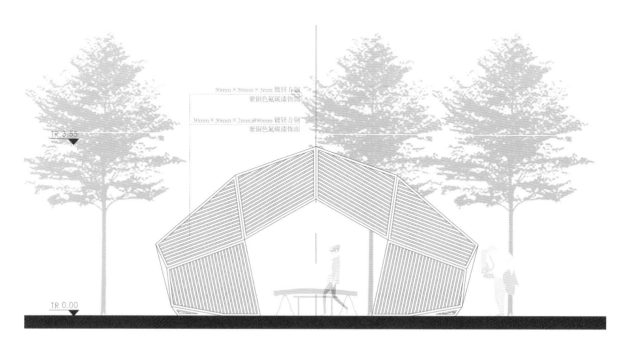

商山阁立面图
Heart pavilion elevation

商山阁效果图
Heart pavilion perspective

置身于一片屋瓦的海洋中看到大地和建筑连为一体
Place yourself in a sea of tiles to feel the land and the building connected together

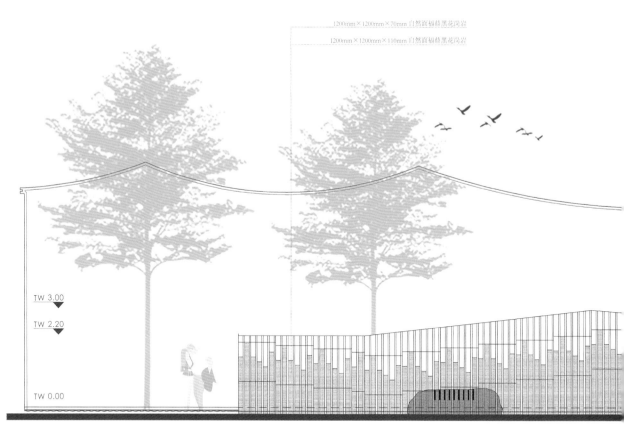

螺峰叠翠景墙展开面
Snail peak pinnacle wall spread out

纯粹之下的场地意境
The artistic conception of the field under pure

瓦构筑的天地
A world made of tiles

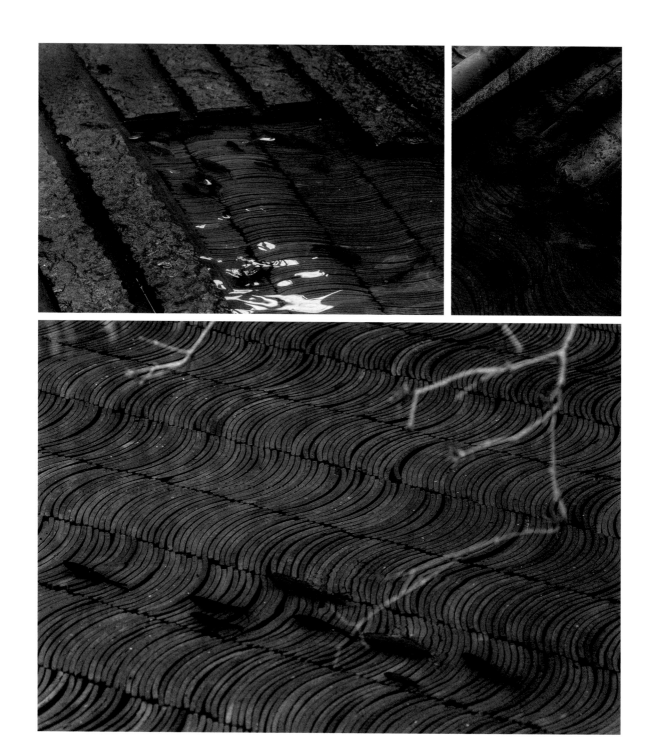

叠叠墨瓦配着薄水一片,红鱼三两结群在水云间掠过
With a thin sheet of water on top of each other, red fish flit between the clouds in twos and threes

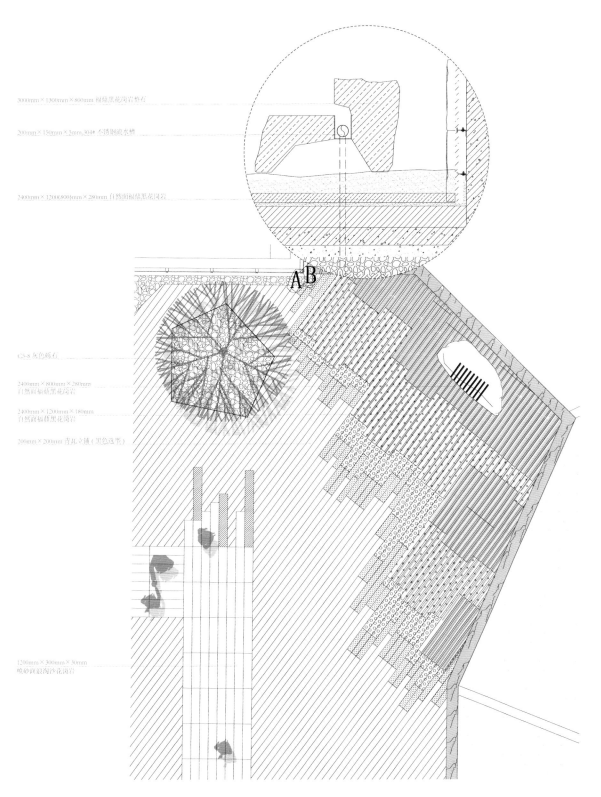

叠翠园平面图
The plan of the emerald green garden

意境的营造成为前院最大的设计亮点
The creation of artistic conception becomes the biggest design highlight of the front yard

景观的微气候综合体 | 福州云影半岛
The Microclimate Complex Of The Landscape
Fuzhou Peninsula Cloud Club

地点 | 福建省福州市
规模 | 7200平方米
业主 | 金辉集团华南区域公司

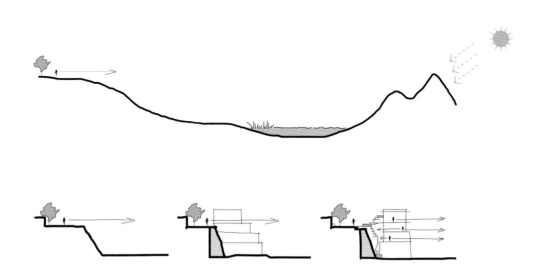

潜层呼吸
| 一条连接地下空间的"峡谷"

半埋在地下的建筑,不能接触充足的光和空气,如果创造一条"峡谷"能够将光导入地下,长长的空间因为空气对流形成了风,峡谷的高差结合水流创造了瀑布,建筑的正形态和峡谷的负形态互为因果,景观和建筑就密不可分了。会所建筑地处福州,风、光、和水创造了适应当地气候的环境空间,炎热的夏季风被导入峡谷进而被流水降温,最炎热时峡谷中比外界低8℃形成非常适宜的避暑空间,冬季面向太阳的台地又是理想观景眺望平台,这条"峡谷"是将景观、建筑、气候复合的天然空调系统。

"Gorge" can lead light to underground. In the long space, because of the air convection, the wind forms. Height difference in the gorge in combination with the water flow creates the waterfall. Positive form of the building and negative form of the gorge interact as both cause and effect. Landscape and architecture are inseparable. The building of the clubhouse is located in Fuzhou. The wind, light, and water create an environmental space adapted to the local climate. The hot summer monsoon is introduced into the gorge and then cooled by the flowing water. At the hottest time, temperature in the gorge is 8 degrees lower than the outside to form a very suitable summer space. The platform facing the sun in winter is also an ideal viewing platform. This "gorge" is a natural air conditioning system that combines landscape, architecture and climate.

原始场地
Original site photos

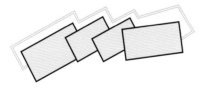

叠级院落
Fold class compound

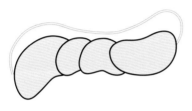

山间云影
The mountain cloud

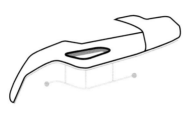
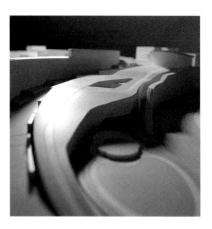

消解高差的方案比较
Comparison of schemes to eliminate height difference

江上云舟
River YunZhou

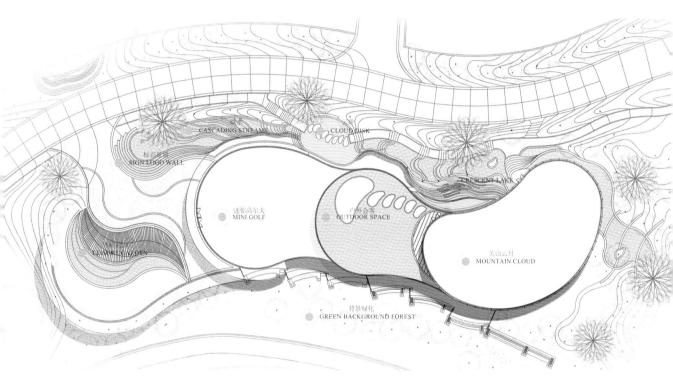

建筑景观的复合形态平面图
A composite shape plan of the architectural landscape

建筑的正形态和峡谷的负形态
The positive form of the building and the negative form of the canyon

汀步的形式模拟建筑的形态
The form of stepping stone inspired from the architectural form

"canyon" introduces light into underground buildings, and water flows through the "canyon" to form air convection, thus creating a microclimate complex.

气候设计的综合体。风从峡谷而来
Complex of climate design on the left page. The wind blew from the canyon

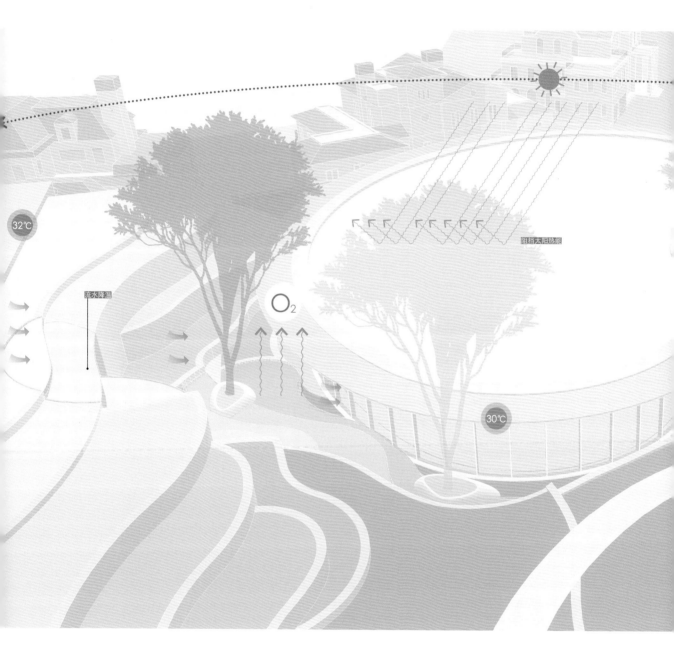

气候峡谷——高差和山地是气候景观的创新重点
Elevation difference and mountain area are the innovation focus of climate landscape

跌水结构分析
Analysis of falling water structure

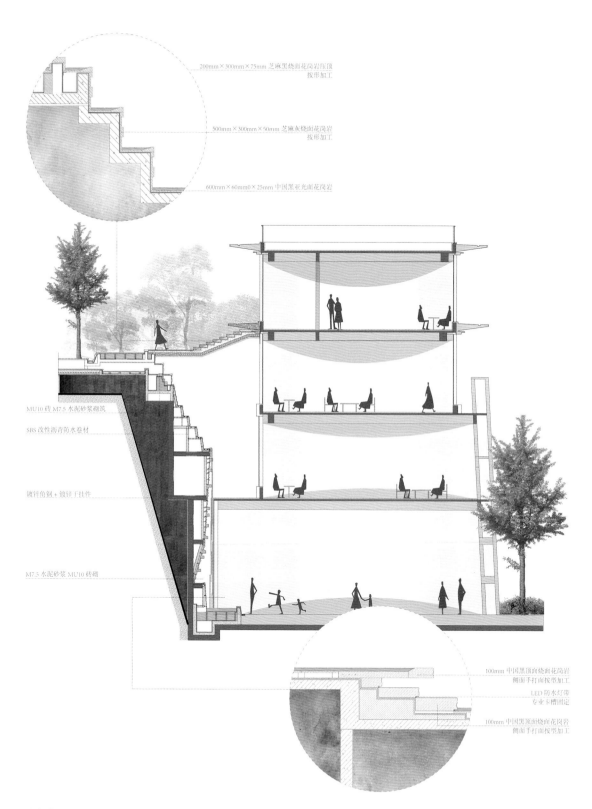

"峡谷"与建筑的关系剖面图
A sectional view of the relationship between the canyon and architecture

景观和建筑的一体化设计
Integrated design of landscape and architecture

一条连接地下空间的"峡谷"
A 'canyon' connecting underground Spaces

跌水剖面图
Section

长长的空间因为空气对流形成了风
The long space is formed by the convection of air into the wind

一条连接地下空间的"峡谷"
A 'canyon' connecting underground Spaces

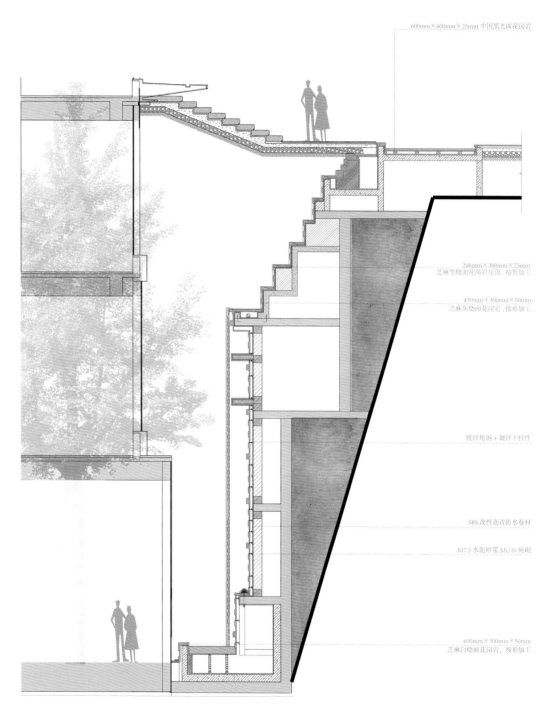

跌水剖面图
Waterscape profile

在高差上创造多样的流水形态
Create various flow patterns on the height difference

微气候的复合设计 | 福州溪溪里
The Composite Design Of Microclimate
FuZhou

地点 | 福建省
规模 | 9.4万平方米
业主 | 金辉集团华南区域公司

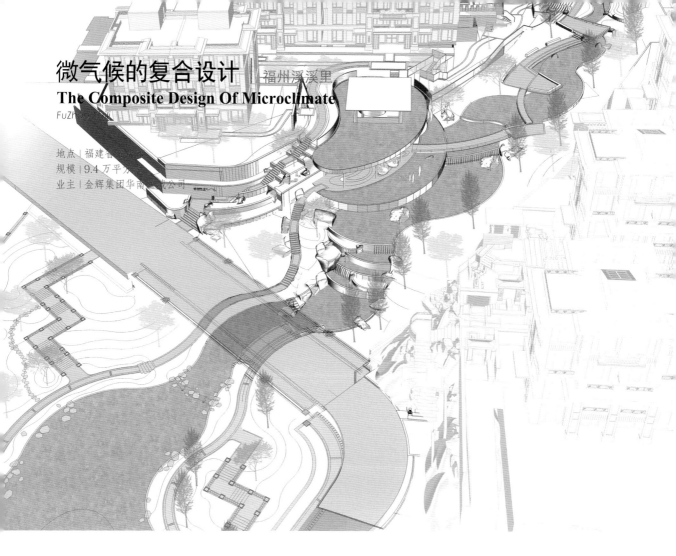

山地微气候理水
一条排洪沟的景观蜕变

排洪沟,这一市政的基础设施,是否有机会变成生活的空间并包容多样的生活方式?

场地原为周边山体"泄洪沟",一条高差十数米的沟渠将两边地块隔开,很长一段时间里作为市政基础设施的"排洪沟"只具有单纯的泄洪功能,而自然状态的溪流其实是结合景观、功能一体的复合形式,设计结合周边地势的特点,以水的微气候特征为主题,对周边建筑进行空间创新的设计,从各个层次联通水系,让水以各种姿态和空间结合。水从高处流下,转换成涌泉、细流、静水、瀑布、水帘、水雾,当这些水的姿态和各种空间形式结合时就形成了微气候空间。人在山地行进,身体姿势随地势变化,拾级而上停留转折间仰头、俯瞰、回望,身体的动作感受在空间中体验贴更细微的温度、体感、听觉的变化,自然山水的意受便不觉浮于心间。

Does the drainage ditch, the municipal infrastructure, have the opportunity to become a living space and accommodate diverse lifestyles?
Area belonged to the surrounding mountain flood discharge trench, a height of ten meters to separate plots on both sides of channels, for a long time as a municipal infrastructure "having ditch" only has the function of pure water, the natural state of the stream is combined with the landscape of composite form, function, design according to the characteristics of the surrounding terrain, with micro climate characteristics of water as the theme, to the design of architectural space surrounding the innovation, from all levels unicom drainage, let water to all kinds of gestures and space combination. The water flows down from the height and transforms into springs, streams, still water, waterfalls, water curtains and water mist. When these water postures are combined with various spatial forms, microclimate space is formed. When people travel in the mountains, their body posture changes with the terrain, and they ascend the stairs to stay at the turning point, looking up, overlooking and looking back. Their body movements and feelings are sensitive to the more subtle changes in temperature, body sense and hearing in the space, so they are imperceptible to the mind of natural landscape.

NO.14 福州溪里

LANDSCAPE DESIGN ABOUT XIXILI

2018 FU ZHOU

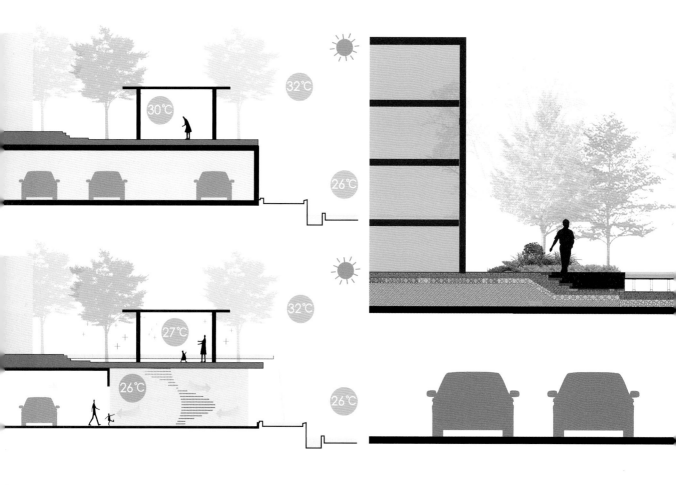

气候设计要求建筑景观一体化
Climate design requires the integration design of architecture and landscape

透过瀑布看风景,营造一处溪流之下的洞穴空间
The view through the waterfall creating a cave space beneath the stream

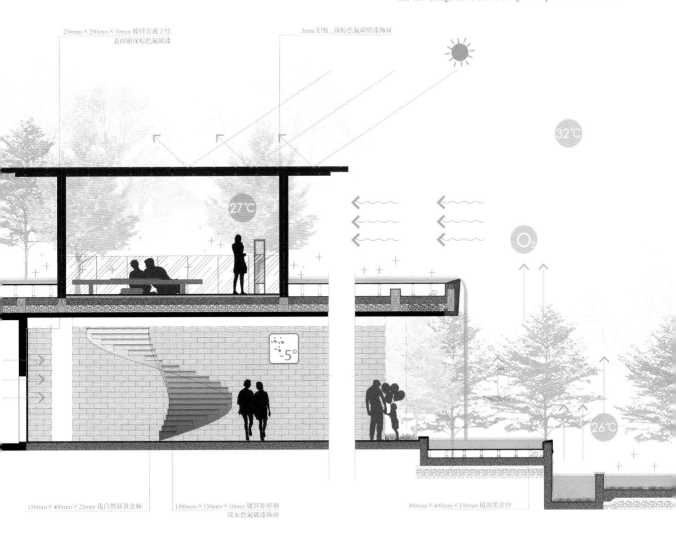

瀑布水亭外观
The waterfall pavilion

景观水冷系统
Landscape water cooling system

363

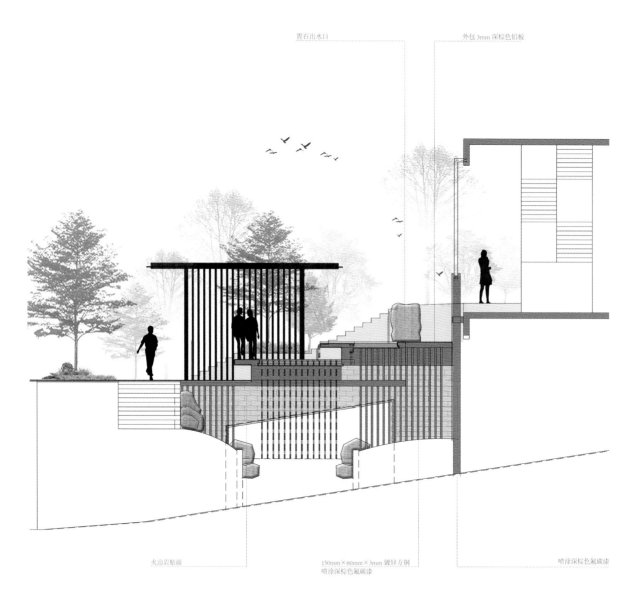

利用高差进行水景设计,创造不同的空间体验
Create unique spatial experiences by combing the design of level and water features

溪水的源头
Stream source

多样的水的姿态
Various water postures

后记 让景观更好地为人服务

邓刚 / 水石设计董事长 创始合伙人

今年是我们的上海水石景观环境设计公司成立 20 周年。水石景观前 10 年的设计实践，基本围绕项目机会，进行多类型、多方向的探索，项目涉及城市公园、居住地产、办公、商业、景观规划等多类型，类型技术能力均质化，项目规模中小型，机构发展处于摸索阶段。近 10 年，设计类型明显集中于地产领域，并同步在城市公园、城市再生、复合型景观领域开始了多元化积累。发展过程中，其设计观念和价值取向更为清晰，设计能力更为综合，项目效果呈现更有品质。既要适应市场需求，也能脱开来思考与服务于设计的本质诉求，这是水石景观实践中一直努力实现的状态。

一是注重功能化与情景感的平衡。在当代的地产设计领域，很多景观设计被狭义地约束在地产展示区、样板区，甚至有设计师开玩笑称，中国产生了一种独特的设计类型——展示区！众多地产设计机构追逐展示区的标志性、风格化，突显极致的情景化，设计要素集中于效果、成本、速度的综合平衡；这必然带来景观设计思维的单一化和对设计能力成长的抑制作用。我们不反对情景化，但是，会更关注功能化；基于功能性的情景化，会更符合设计的本质需求。为此，我们通过研究人体工效、人的行为模式、客户使用需求、专项类型研究等，力争建立基于人使用和感知为基础的设计方法、工具、策略，以及在设计中充分实践。

二是强调科学化与感性的平衡。相比于建筑设计，景观设计规范性弱，行业门槛低；尤其是狭义的景观设计，更容易被简单分解为室外空间设计、硬质景观、软景设计等；过于随意的总图布局，过于感性的风格化倾向，都是当代景观设计中的常态。而水石景观会研发专业设计工具辅助设计，增加设计中的理性成分，数据让设计更有依据，研发积累让设计更为成熟。水石的设计表达强调还原度，通过数据建模、实体模型、表现图互相印证设计效果，让局部精细化建模表达，充分呈现设计细节，让设计更容易被理解和阅读，方便施工。我们强调形态既要表达情感和美学，设计也需要有逻辑；这些都促进了景观设计的水平提升。

三是探索建筑、规划、景观的一体化呈现。依托水石在城市再生、地产设计领域的项目机会，水石在深度实践跨专业的设计无缝对接。我们会力争以更开阔的视野去理解景观，从规划、策划、建筑、室内的多元角度，看待景观专业与之的互动，甚至是以广义景观思维，整合其他专业；其次，更注重知识结构的综合性，即使在景观专业领域，功能化、情景感、经济性、施工工艺等等，都是我们的综合范畴；这样的努力也开始拉开了水石与其他设计机构的能力向度，让基于专业化的综合能力成为服务能力提升的抓手。

越是在景观设计领域的广度与深度实践，越是深刻理解设计的本质。无论是功能化与情景感的平衡，还是设计中科学化与感性的平衡，或者我们努力探索的建筑、规划、景观一体化，其围绕的核心就是让景观设计更好地为人服务！水石设计机构的定位，也是以综合设计服务为主要能力的设计平台。本书中 14 个多姿多彩的实践案例，既是对过往的 20 年之精彩记录；也是我们让景观设计更好地为人服务观念的探索与实践。作为水石设计的创始人之一，我尤其欣喜地认识到，本书所有呈现的案例，都是水石 80 后设计团队作品！

今年 3 月份，在法国戛纳，我作为项目总负责人，代表长春水文化生态园的水石设计团队荣获了 MIPIM AWARDS 入围奖，我想这一定程度上，也是同业对水石所努力方向及设计实践一定程度的认同吧。在接下来的发展中，我们会继续遵循让设计更好地为人服务的宗旨，让设计创造价值，相信水石会有更多出色的实践与积累！期待水石的下一个 20 年！

POSTSCRIPT
LET THE LANDSCAPE DESIGN SERVE PEOPLE BETTER

Gang DENG / CHAIRMAN& FOUNDING PARTNER OF SHUISHI

This year, Shanghai SHUISHI Landscape Environment Design Co., Ltd. will celebrate her 20th anniversary, and the first 10 years of her design practice was basically centered on multi-type and multi-directional explorations in various projects, including city parks, residential properties, office, commerce and landscape planning, etc. . At this exploratory stage of development, types of design and technical capabilities were homogeneous, and the projects were basically small or medium in size. Over the recent 10 years, SHUISHI Landscape has focused her designs more on real estate and expanded to projects of city parks, urban renewal and composite landscapes. In her development, the design concept and value orientation are being better defined, the design abilities more comprehensive, and the project effect better qualified. While adapting to the market demand, SHUISHI has to take a step back to think and serve the essence of design, the state that SHUISHI Landscape has always been trying to achieve in practice.

Firstly, the balance between function and situation. In contemporary real estate design, many landscape designs are limited to property display areas and sample areas, and some designers even joke that China has created a unique type, display area! Many real estate design institutes are addicted to symbolic and stylistic display areas, highlighting extremely situational design. The design elements are mainly for the comprehensive balance of effect, cost and speed. This will inevitably simplify the landscape design thinking and harm the improvement of design ability. We are not against situationalization, but we prefer to pay more attention to functions; a function-based situationalization will better fit the essential needs of design. We therefore study ergonomics, patterns of human behavior, user requirements, special type research, etc., try to establish design methods, tools and strategies on the basis of human use and perception, and fully practice them in design activities.

Secondly, the balance between sense and sensibility. Compared with architectural design, the design of landscape has less regulatory requirements and seemingly everyone can do it. In the narrow sense of landscape design, in particular, it's simply divided into outdoor space design, hard landscape, soft landscape design, etc. Casual and free general layouts and emotional styles are quite common in contemporary landscape design. SHUISHI Landscape will develop professional design tools to increase sensible elements in the design. The data make the design more reliable, and the research and development ensure mature designs. And SHUISHI design expression emphasizes revivification, and confirms the design effect through data modeling, physical model and rendering graphics; and the refined regional modeling expression fully presents the design details to make it easier to read, understand and execute the design. We emphasize that the form should express emotion and aesthetics, and the design be logical. And all these have improved the level of landscape design.

Thirdly, the integration of architecture, planning and landscape. SHUISHI makes use of her project opportunity in urban renewal and real estate design, and experiments seamless integration of cross-specialty designs. We will strive to assume a broader perspective to understand landscape, view the interaction between landscape and planning, scheming, architecture, interior design from the latter aspects' perspective, and even integrate the other specialties under the broader thinking of landscape design. And we will pay more attention to a comprehensive knowledge structure, and subject functionalization, situationality, economic performance and construction techniques under landscape specialty. Such efforts have also begun to distinguish SHUISHI from other design institutes in terms of ability and made the professional integration ability a booster to improve service capabilities.

The more extensive and in-depth is the practice in landscape design, the more deeply we understand the nature of design. For the balance between function and situation, the balance between sense and sensibility, and the integration of architecture, planning and landscape, the core is to make landscape design better serve the people! SHUISHI is a design platform that provides integrated design service. This book contains 14 practical cases of landscape designs. It's a wonderful record of the past 20 years' development as well as our exploration and practice to make landscape design better serve the people. As a co-founder of SHUISHI , I am delighted to realize that all the cases in this book are the works of SHUISHI 's designers who were born after the 1980s!

This March, I, as the chief project manager and on behalf of W&R designers of Changchun Culture of Water Ecology Park, received in Cannes the finalist award of the MIPIM AWARDS. I think this award, to some extent, represents the recognition by the landscape design industry of the efforts and design practice of SHUISHI. In the future development, we will continue to follow the tenet of making design better serve the people, and let the design create value. I believe SHUISHI Design will have more excellent practice and harvest! We look forward to the next 20 years of SHUISHI !

01 —
长春水文化生态园
Culture Of Water Ecology Park ChangChun
地点：长春市

设计团队：
张淞豪 王慧源 黄建军 廖勐乙 何鑫 顾婧 张进省
李花文 曾杰烽 蒋婷婷 焦翔 裴杰 李鸿基 谢金倍
褚昌浩 杜瑞文 赵俊 孙震 李建军 徐光耀 杨智博
吴嘉鑫 邵莹 赵凯 金光伟 黄晓晨 夏伟龙 李珊婷
金戈 李斌 潘运泓 王涌臣 赵延伟 徐晋巍 张帅
陈浩 方雁容 贾京涛 朱泉林 陈伟忠 曹旭 叶田
杜侠伟 李一博 徐君韦 孙海校 余钢 胡颉 徐梦茜

02 —
西安白鹿原西坡树梢漫步栈道
Rambling Plank Road At The Treetop Of Western Slopes Of Bailuyuan XiAn
地点：西安市

设计团队：
张淞豪 黄建军 王慧源 李洪基 谢金倍 黄玥晴
江敏圣 王浩 王月 任维护 蒋春丽 侯亚凤 辛敬芳
刘亚婷 钟琴 贺鑫 刘佳 王文珍

03 —
上海田林路改造
ShangHai Tianlin Road Renovation
地点：上海市

设计团队：
张淞豪 周良杰 高崧 董怡嘉 刘璞
张宇鹏 徐阳 程子轩 朱俊杰 杨阳

08 —
南昌红土遗址公园
Red Earth Heritage Park Nanchang
地点：南昌市

设计团队：
孙翀 石力 赵晓东阳 石岱 向左明 陈宇奇 龙勇
田卓颖 祁峰 陈佳毅 付冬冬 刘文静 金靖

09 —
三亚 JW 万豪酒店
JW Marriott Hotel Sanya
地点：三亚市

设计团队：
石力 马纯皓 何彬 王怀进 张进省 李牧临 郭豪特

10 —
郑州长安古寨
Zhengzhou Chang'an Ancient Village
地点：临沂市

设计团队：
石力 王燕 刘炜若 谭薛巍 吴静娴 姜孙鼎
翟丹 张扬徐 王文珍

04 —
西安万科理想城
Ideal City XiAn
地点：西安市

设计团队：
张淞豪 王慧源 何鑫 杜瑞文
刘建军 顾婧 李花文 梁巧会
廖勐乙 蒋婷婷 李鸿基 黄玥晴

05 —
南昌博览城绿带公园
Guobo Green Park In NanChang
地点：南昌市

设计团队：
张淞豪 王慧源 何鑫 曾杰烽
刘建军 顾婧 王月 李花文
梁巧会 张进省 徐捷

06 —
南昌博览城庆典公园
Celebration Park In NanChang
地点：南昌市

设计团队：
张淞豪 王慧源 何鑫 曾杰烽
杜瑞文 王月 李花文 梁巧会
裴洁 李静

07 —
南昌安南小镇玻璃花房
Annan Town Glass Greenhouse Nanchang
地点：南昌市

设计团队：
石力 孙翀 赵晓东阳 石岱 陈宇奇 祁峰
陈佳毅

11 —
临沂生命织树
Tree of Life in LinYi
地点：临沂市

设计团队：
张永良 刘国泰 曹天蟾 杨莉 吴琪

12 —
昆明樾府
Kunming Oriental Mansion
地点：昆明市

设计团队：
石力 张淞豪 黄建军 杨娜 谢云建
石岱 王文泉 唐继平 蒋立薪
王文珍 高树桃 李巧

13 —
福州云影半岛
Fuzhou Peninsula Cloud Club
地点：福州市

设计团队：
石力 赵晓东阳 付胖 田卓颖 祁峰 金婧
付冬冬 刘文静 陈佳毅 王文珍 王泽玉

14 —
福州溪溪里
FuZhou Xixili
地点：福州市

设计团队：
石力 杨娜 黄崇伟 刘晓玲 邓筱敏 付胖
陈俊 祁锋 王月 张进省 陈啸 张晓辉
温婧 王文珍 高树桃

图书在版编目（CIP）数据

会呼吸的景观 / 水石设计著 . ——上海：同济大学出版社，2019.4
ISBN 978-7-5608-8521-6

Ⅰ．①会… Ⅱ．①水… Ⅲ．①景观设计 – 作品集 – 中国 – 现代 Ⅳ．① TU983

中国版本图书馆 CIP 数据核字 (2019) 第 065658 号

出品：
水石设计

策划：
张淞豪　　石力

编委：
张淞豪　　石力　　张永良　　孙翀　　周良杰　　黄建军　　王慧源　　张妍钰

技术支持：
石岱　何鑫　杜瑞文　廖勐乙　蒋婷婷　曾杰锋　顾婧　李洪基　李雨倩　张玉　王鸿晗　宋丽丽　杨雨金　杨元兵　刘亚　岳森　葛巍　王文泉　赵晓东阳　陈宇奇　龙勇　田卓颖　郭豪特　高崧　张宇鹏　程子轩　徐阳　辛静芳　钟琴　刘亚婷　贺鑫　王晓慧　朱明龙　王佳娣　苍静　江灵敏　宋建勋　全轩震

书名：会呼吸的景观

责任编辑：荆华　　责任校对：徐春莲　　装帧设计：石岱

出版发行　同济大学出版社　www.tongjipress.com.cn
　　　　　（地址：上海市四平路1239号　邮编：200092　电话：021-65985622）
经　销　全国各地新华书店
印　刷　上海雅昌艺术印刷有限公司
开　本　889mm×1194mm　1/16
印　张　23.5
字　数　752 000
版　次　2019 年 4 月第 1 版　2019 年 4 月第 1 次印刷
书　号　ISBN 978-7-5608-8521-6
定　价　268.00 元

本书若有印装质量问题，请向本社发行部调换　版权所有　侵权必究